INTRODUCTION

I don't remember the exact moment I first heard the phrase "artificial intelligence." It might have been a headline, a classroom lecture, or a passing conversation. But I remember the feeling—curiosity tinged with wonder, as if someone had cracked open a window into the future. Machines that could think? The idea clung to me like static electricity.

Years have passed since that first spark of intrigue, and in that time, AI has gone from science fiction to everyday fact. It recommends what we watch, replies to our emails, writes stories, paints images, and diagnoses diseases. It whispers into our pockets, answers before we ask, and finishes our sentences. And somewhere in this quiet revolution, I found myself asking:

What does it mean to be the "I" in AI?

This book is a reflection of that question. A meditation on identity, agency, emotion, and intelligence in a world where machines are beginning to mirror us—not just in tasks, but in thought and even creativity. I in AI isn't about circuits or code. It's about people. About how we respond when we look into the glowing eyes of our own inventions and see something unsettlingly familiar staring back.

We are overwhelmed by technology—awed by its brilliance, but also quietly anxious. Will it outpace us? Understand us? Replace us? Can it learn? Can it feel? Can it care?

And more urgently: **If machines grow more human, how do we hold on to what makes us human?**

I don't come to this page as a prophet or a programmer. I come as a person—flawed, fascinated, and full of questions. This book doesn't offer answers as much as it opens windows. Each chapter explores a different intersection:

of art and algorithm, empathy and logic, creativity and calculation. It's a journey of observation, reflection, and at times, confrontation.

Because in the end, AI is not the story. We are.

And so, I invite you—not just to read, but to wonder with me. To hold space for nuance, to sit with discomfort, to dream responsibly. Let us explore, together, what intelligence really means when it's no longer ours alone.

Let's begin.

You and I.

And everything in between.

TABLE OF CONTENTS

Chapter 1

The Nature of Intelligence - Human vs.
Machine............................. 1

Chapter 2

The Human-AI Relationship —Friend, Foe, or

Just a Tool?............. 15

Chapter 3

AI at Work — A New Partner or a
Replacement?.............................. 27

Chapter 4

AI in Social Life — The Double-Edged
Sword.................................... 37

Chapter 5

AI and Decision-Making: Who Do We
Trust?.................................... 49

Chapter 6

Ethics in AI: The Dark Side of
Intelligence... 62

Chapter 7

Redefining Humanity: What Does It Me to Be Intelligent?............... 74

Chapter 8

Conclusion: Keeping the 'I' in AI.. 85

Voices on AI... 98

About the Author.................................... 99

References and Reading List.. 100

Acknowledgements................................. 102

'The question is not whether intelligent machines can have any   emotions, but whether machines can be intelligent without any emotions.'

— Marvin Minsky, AI Pioneer

The Nature of Intelligence —

Human vs. Machine

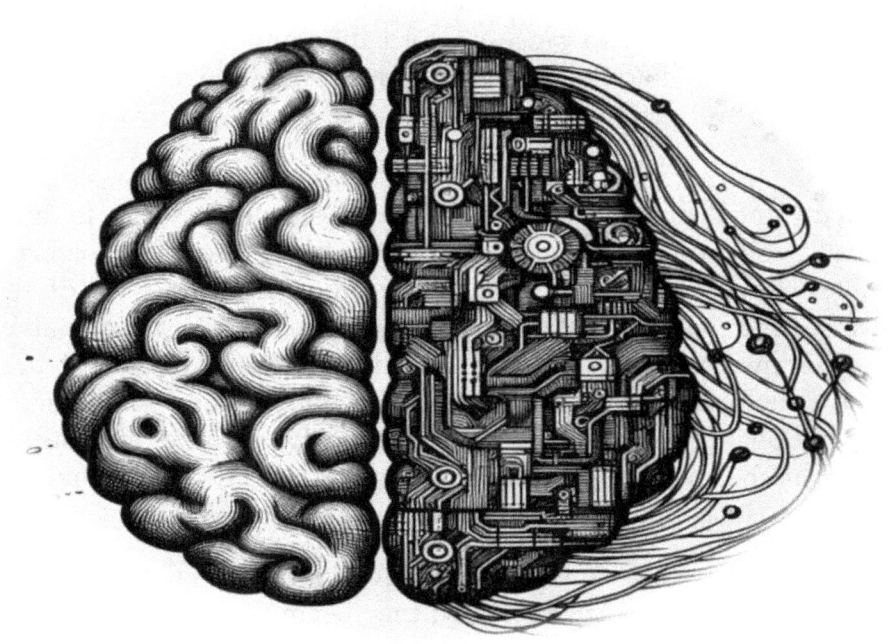

What makes intelligence, well, *intelligent*? Is it simply the ability to learn and solve problems, or is it more than that? Intelligence, as we humans perceive it, is a complex phenomenon. It encompasses not only knowledge but also creativity, intuition, and emotional depth. Artificial Intelligence (AI) introduces a new twist to this ancient concept. Machines can solve problems faster than we ever could, paint images in milliseconds, and even hold conversations that feel human. But can they truly be called intelligent?

Defining Intelligence

To truly explore intelligence, we must unravel a concept that has fascinated humanity for centuries. From Aristotle's contemplations on reason to modern explorations in cognitive science and Artificial Intelligence (AI), intelligence has been studied, debated, and endlessly redefined. But what exactly do we mean when we talk about intelligence? Is it simply the ability to process information and solve problems, or does it go deeper—into the realms of emotion, intuition, and creativity? And how do we distinguish the intelligence of humans from that of machines?

For many, intelligence is synonymous with the mind's capacity to think critically, solve problems, and analyse situations. A student solving a complex math equation, a scientist decoding the mysteries of the universe, or an engineer designing a groundbreaking invention—all of these examples reflect intellectual capability. Yet even in these analytical pursuits, human intelligence is far more complex than raw logic or computational prowess. It is not just about arriving at the right answer but about the process—intuition, curiosity, trial and error, and the moments of imaginative insight that lead to breakthroughs.

For instance, take the scientific method. A scientist does not approach a problem with all the answers ready; they begin with questions. They hypothesize, observe patterns, and adapt their approach based on results.

Machines, too, are designed to analyse data and learn from it, but their "learning" follows a structured algorithmic process. They lack the imaginative leap, the "aha" moment born from curiosity and experience, that often propels human discovery. This distinction underscores the multifaceted nature of human intelligence—it is not just about processing data but about interpreting it in ways that resonate with meaning and purpose.

One of the most defining aspects of human intelligence is its deeply emotional and social nature. Intelligence is not confined to solving puzzles or designing systems; it is also about navigating relationships, understanding others, and forging connections. Picture an artist painting a portrait, not merely to reproduce someone's features but to capture the essence of their personality and emotions. Or consider a teacher who adjusts their methods to inspire a struggling student, demonstrating emotional intelligence that bridges the gap between knowledge and understanding. Human intelligence is profoundly relational, born from our shared experiences and the way we connect with one another.

Machines, on the other hand, excel in replicating aspects of intelligence that are quantifiable: calculations, pattern recognition, and data analysis. They are powerful tools for solving specific problems, such as diagnosing diseases based on medical data or optimizing traffic flow in a city. For example, AI can process thousands of scans to detect early signs of cancer far faster than a human doctor ever could. Yet, this intelligence operates within a narrow framework—it is defined by the boundaries of its programming and training data. While a machine might outperform a doctor in identifying anomalies, it cannot console a worried patient or adapt its approach based on the nuances of the patient's personal story. These limitations highlight the gap between functional efficiency and the human ability to empathize, imagine, and respond to the complexities of real life.

Creativity further illuminates the uniqueness of human intelligence. AI systems like those designed to generate art or compose music are often heralded as examples of machine creativity. They can analyse vast datasets of existing works to produce paintings or songs that mimic specific styles. And yet, the process of creation for humans is rarely just a technical exercise. A poet writes not just to string together beautiful words but to explore their own emotions, to heal from heartbreak, or to share a universal truth. An AI can certainly generate a poem that rhymes and follows grammatical rules, but it

cannot infuse that poem with the lived experiences, vulnerabilities, or aspirations that give human art its soul. Creativity, for humans, is deeply tied to intention, struggle, and the ineffable spark of imagination.

Another key facet of human intelligence lies in intuition—a form of knowing that isn't easily explained by logic alone. Think of a chess player making a seemingly risky move, not because they've calculated every possible outcome, but because they "sense" it's the right choice.

Or consider a firefighter making a split-second decision in a life-threatening situation, guided by instinct honed through experience. These moments of intuition are born from a complex interplay of memory, emotion, and subconscious processing, which cannot be easily replicated by machines.

Moreover, intelligence, for humans, includes the ability to reflect on ourselves and our place in the world. This metacognitive ability—to think about our thinking, to question our motivations, and to strive for self-improvement—is uniquely human. It allows us to engage in philosophical debates, to grapple with existential questions, and to seek meaning beyond the immediate. Machines, no matter how sophisticated, do not engage in this kind of introspection. They do not ponder the implications of their calculations or question the ethics of their programming. These are uniquely human pursuits, rooted in our capacity for self-awareness and moral reasoning.

To better illustrate the distinction between human and machine intelligence, consider the following analogy. Imagine a world-class pianist performing a piece by Beethoven. Every note they play is informed by years of practice, emotional interpretation, and a connection to the music that transcends technical skill. Now imagine a player piano, perfectly replicating that same piece with mechanical precision. While the performance may sound flawless, it lacks the depth, nuance, and expression that make the pianist's interpretation so profoundly moving. Similarly, while AI can replicate certain aspects of intelligence, it cannot replicate the essence of what makes intelligence uniquely human.

In summary, intelligence is not a single, easily defined attribute. For humans, it encompasses logical reasoning, emotional depth, creativity, intuition, and

the ability to connect with others in meaningful ways. Machines, impressive as they are, operate within the boundaries of their design, excelling in tasks that are structured, logical, and data-driven. They are tools that enhance and extend human capabilities but do not replace the richness and complexity of human thought and emotion. Understanding these distinctions is not just an academic exercise; it is essential for navigating the age of AI with wisdom and intention, ensuring that we celebrate and preserve the qualities that make us human.

Human Intelligence: More Than Just Reasoning

Human intelligence is more than just the capacity for reasoning—it is the intricate interplay of logic, emotion, intuition, and connection that defines how we perceive and interact with the world. Among these elements, emotions stand out as a cornerstone of human intelligence, shaping our decisions, relationships, and even our creativity. They add depth and meaning to our experiences, guiding us in ways that pure logic never could. Emotions enrich our lives and serve as a compass in moments when reasoning alone cannot provide clarity. To truly understand the distinction between human intelligence and artificial intelligence, emotions provide an illuminating starting point.

Imagine witnessing a close friend achieve a lifelong dream—graduating from a challenging program, landing a long-coveted job, or simply overcoming a personal hurdle that once seemed insurmountable. As they beam with pride, you instinctively smile, feeling their joy resonate within you. There is no deliberate calculation, no formula that dictates your reaction. Your shared history, the memories you've built together, and your emotional connection create an almost automatic response: you share their happiness as though it were your own. Now, imagine the opposite scenario—your friend is in sorrow, perhaps after a devastating loss or a moment of deep personal struggle. You offer words of comfort, not as a premeditated act but as an expression of empathy born from your own experiences with grief. You don't need to analyse the situation logically to know what to say or do; your own lived experiences, combined with your emotional awareness, guide you.

This profound capacity for empathy—the ability to understand and share the feelings of another person—is central to human intelligence. It allows us to form deep, meaningful connections and to navigate complex social landscapes with sensitivity and care. Empathy arises not just from cognitive recognition of another's emotions but from our emotional depth, memories, and lived experiences. It is a uniquely human phenomenon, woven into the fabric of who we are. Machines, no matter how advanced, lack this capacity for true emotional resonance. Even the most sophisticated artificial intelligence systems remain confined to the realm of simulation.

Consider how AI attempts to mimic empathy. Sentiment analysis enables machines to identify the tone of a message, distinguishing between joy, sadness, anger, or frustration.

Chatbots and virtual assistants can respond to phrases like "I'm feeling down" with programmed replies such as "I'm sorry to hear that. Is there anything I can do to help?" On the surface, these responses seem empathetic—they acknowledge the user's emotions and attempt to provide comfort. But beneath the surface, they are mere predictions generated by algorithms trained on vast datasets of human interactions. An AI does not truly "feel sorry"; it cannot understand the sorrow it is addressing. It doesn't experience the complex, visceral nature of human emotions. It simply recognizes patterns in text and generates a response it deems appropriate based on statistical probabilities.

To illustrate this difference, let's compare a human counsellor to an AI chatbot designed for mental health support. A human counsellor brings not just professional training but also their own life experiences and emotional depth to each interaction. When they listen to a patient describe feelings of loss, they do so with an understanding shaped by their own experiences with grief or moments of struggle. Their responses are infused with compassion, shaped by a genuine desire to support another person in need. An AI chatbot, on the other hand, might analyse the same interaction and offer a reply that appears thoughtful or supportive. Yet, this reply is devoid of intention or understanding. The AI does not "care" for the person it is assisting; it has no context beyond the data it has been trained on and no capacity to connect on a personal level.

The inability of machines to experience emotion also limits their ability to make decisions that require emotional intelligence. For example, a leader facing a difficult choice might weigh not just the logical outcomes but also the emotional impact on their team. They might decide to deliver a challenging message with care and sensitivity, knowing how it will be received and wanting to minimize harm. A machine in the same scenario would lack this nuanced awareness. It might choose the most efficient solution without considering its emotional repercussions, because efficiency—not emotional well-being—is the metric it is designed to prioritize.

Creativity is another area where emotions play a critical role. When an artist paints or a poet writes, their work is often an expression of their inner emotional world—joy, despair, love, or longing. These emotions provide the spark of inspiration and guide the creative process in ways that are deeply personal and intuitive. While AI can generate poems or paintings, these creations lack the lived experiences and emotional narratives that imbue human art with meaning.

An AI might produce a beautiful landscape painting, but it cannot capture the sense of awe a human artist feels standing before that landscape at sunrise. It might generate a poignant poem, but it does not feel the heartbreak or hope that inspired the words.

This distinction becomes especially evident in moments of vulnerability. Imagine a parent comforting their child after a nightmare. The parent doesn't just offer words of reassurance; they hold the child, stroke their hair, and share stories of their own childhood fears to help the child feel safe. These acts of comfort are not calculated; they flow naturally from the parent's empathy and love. A robot designed for childcare might mimic some of these actions—offering soothing words or playing calming music—but it does so without intention or understanding. It cannot feel the child's fear or respond with the depth of care that arises from emotional connection.

In recognizing the limitations of artificial intelligence, we gain a deeper appreciation for what makes human intelligence so extraordinary. Emotions are not a weakness; they are a strength.

They guide our decisions, enhance our creativity, and deepen our connections with others. They allow us to navigate the complexities of life with compassion, resilience, and adaptability. Machines, for all their computational power, operate in a world of patterns and probabilities. They excel at tasks that require efficiency and precision but falter when faced with the nuances of human experience.

As we continue to develop AI and integrate it into our lives, it is essential to remember the irreplaceable role of emotions in human intelligence. Emotions are not just a byproduct of our humanity; they are at its core. They remind us that intelligence is not merely about solving problems or achieving goals—it is about understanding, connecting, and creating meaning. While AI may augment our abilities and reshape our world, the essence of human intelligence—rooted in emotion, empathy, and lived experience—will remain uniquely and beautifully ours.

The Role of Creativity and Imagination

Another hallmark of human intelligence is creativity—the ability to generate novel ideas, see connections where others see none, and imagine possibilities beyond what is known. Consider a painter staring at a blank canvas. They might draw inspiration from their childhood, a recent heartbreak, or the beauty of a sunset. Every brushstroke is infused with emotion, memory, and meaning.

Now compare this to an AI system trained to create art. AI-generated images can be stunning, even breathtaking, but they are not born of imagination. Instead, the machine analyses millions of existing artworks, identifies patterns and styles, and combines them to produce something new. While the output might look creative, it lacks the emotional depth and originality that comes from a uniquely human perspective.

Take the example of OpenAI's DALL-E, an AI capable of generating images from text descriptions. It can create a picture of "a cat sitting on a moonlit windowsill, gazing at the stars," but this is not the result of inspiration or imagination. It is simply the outcome of statistical probabilities derived from its training data.

Intuition: The Unspoken Gift

Intuition is another defining aspect of human intelligence—an almost inexplicable ability to know something without relying on conscious reasoning. Imagine you're driving on a foggy road and suddenly feel the need to slow down, only to discover an obstacle moments later. You didn't logically analyse the situation; you simply *felt* something was off. This gut feeling often stems from subconscious patterns we've learned over time, combined with an awareness that goes beyond data points.

Machines, on the other hand, rely entirely on data to make predictions or decisions. An AI-powered self-driving car might detect the same obstacle using sensors and algorithms, but this is a calculated response based on programming, not intuition.

While incredibly effective in structured environments, machines struggle in situations that require instinct or adaptability to ambiguous circumstances.

Social Interaction and Connection

Human intelligence also thrives in social environments. We are wired to form relationships, read subtle cues, and adapt our behaviour based on the emotions and needs of others. For example, during a heartfelt conversation, you might notice your friend's slight hesitation or a change in their tone of voice, signalling that they're holding back something important. You adjust your approach, encouraging them to share more openly.

AI, in contrast, processes interactions as input-output exchanges. While conversational AI like ChatGPT or customer service bots can provide coherent and even empathetic responses, they lack the deeper understanding that underpins true social intelligence. A machine cannot pick up on the hidden layers of meaning in a pause or a sigh—it can only respond to the explicit data it is given.

The Enigma of Emotional Resonance

Ultimately, what sets human intelligence apart is its emotional resonance—the ability to connect, to feel, and to imbue every decision or creation with meaning. Machines, no matter how advanced, operate within the boundaries of their programming. They can simulate understanding and even outpace humans in specific tasks, but true intelligence involves far more than raw computational power.

Imagine a poet crafting verses about love and loss. Each word reflects their unique perspective, shaped by personal experiences and emotions. An AI might generate a poem that mimics the structure and language of great poets, but it doesn't understand love or loss—it merely assembles patterns based on its training.

This is the essence of intelligence: it is not just the capacity to solve problems or process information but the ability to experience, create, and connect.

It is the depth of human experience, rich with emotion and imagination, that makes intelligence truly remarkable—and it is what no machine can replicate.

The Rise of Machine Intelligence

Artificial intelligence had humble beginnings. The earliest computers were nothing more than calculators, designed to perform tasks like arithmetic faster than any human could. But these machines evolved. By the 1950s, researchers began to dream of creating systems that could think and learn. Alan Turing, often called the father of AI, famously posed the question: *Can machines think?*

One of the first AI breakthroughs came in the form of chess. In 1997, IBM's Deep Blue defeated Garry Kasparov, a world chess champion. This moment marked a significant leap in machine intelligence. Yet, even as Deep Blue won, it wasn't truly *thinking*. The machine calculated millions of potential moves in seconds, relying on sheer computational power.

Compare that to Kasparov's approach—drawing on intuition, experience, and an almost emotional connection to the game. While Deep Blue was a remarkable tool, it lacked the human spark that defines intelligence.

Human Intuition vs. Machine Pattern Recognition

A fundamental difference between human and machine intelligence lies in how they process information. Humans often rely on intuition—a gut feeling that isn't always logical but is remarkably effective. Picture this: You're walking through a park and sense that it's about to rain. You haven't checked the weather forecast, but a combination of darkening clouds, a drop in temperature, and subtle changes in the air give you this intuitive insight. Machines, on the other hand, would analyse weather data, atmospheric pressure, and precipitation patterns to reach the same conclusion.

This raises the question: which form of intelligence is more effective? The answer isn't straightforward. Machines excel at tasks that involve vast amounts of data. For example, AI systems in healthcare can analyse thousands of medical records to detect early signs of disease.

A human doctor might miss these patterns simply because they don't have the capacity to process as much information. But when it comes to bedside manner—the empathy and reassurance that can make all the difference to a patient—machines fall short.

Creativity: The Human Edge

Can machines be creative? It's a question that has sparked endless debate. AI models can generate stunning artwork, compose music, and even write stories. Take, for instance, the AI-generated portrait *Edmond de Belamy,* which sold for $432,500 at auction. Critics and enthusiasts alike marvelled at the machine's ability to create. But was it true creativity?

When a human creates something, whether it's a painting or a piece of music, it often reflects their emotions, experiences, and unique perspective. An artist pouring their soul onto a canvas isn't just combining colours—they're telling a story. AI, in contrast, generates content by analysing patterns and mimicking what it has been trained on. It doesn't feel joy, sorrow, or inspiration. Its creations, while impressive, lack the depth and authenticity that come from lived experience.

The Philosophical Question: What Makes Intelligence Unique?

As we compare human and machine intelligence, we encounter a philosophical challenge. If machines can outperform us in tasks like problem-solving and data analysis, does that make them superior? Or does the lack of consciousness, emotion, and free will make them fundamentally different?

Imagine a marathon runner competing against a self-driving car. The car, equipped with advanced sensors and algorithms, will undoubtedly reach the finish line faster. But does that make it the *better* runner?

The human runner brings something to the race that the car never can— a story, a journey, a sense of triumph. Intelligence is about more than results; it's about meaning.

Real-Life Applications and Challenges

AI is transforming industries around the globe. In agriculture, AI-powered drones monitor crops and optimize yields. In education, AI tutors personalize learning experiences for students. In the arts, AI tools assist musicians and filmmakers in creating groundbreaking works. But with these advancements come ethical dilemmas.

For example, when AI systems are used in hiring, they may inadvertently reinforce biases present in their training data. This highlights a critical flaw in machine intelligence: while it can process data, it doesn't understand context. A human recruiter, in contrast, might recognize the nuances of a candidate's experience that an algorithm overlooks.

Human and AI: A Collaboration, not a Competition

As we move forward, the relationship between human and machine intelligence doesn't have to be adversarial. Instead, it can be collaborative. Think of AI as a tool that enhances our abilities rather than replacing them. When a human doctor works alongside an AI diagnostic tool, the combination of empathy and data-driven insight leads to better outcomes. When an artist uses AI to experiment with new styles, the result is a fusion of technology and creativity.

The question isn't whether machines will surpass humans—it's how we can work together to create a future where intelligence, in all its forms, serves the greater good.

Conclusion

As we arrive at the conclusion of this chapter, we find ourselves at a pivotal moment—a moment where the exploration of intelligence transitions from analysis to reflection. We've delved deeply into the multifaceted nature of intelligence, dissecting its definitions, tracing its evolution, and examining its manifestations in both humans and machines. While the advancements of Artificial Intelligence (AI) are nothing short of remarkable, they illuminate a profound and undeniable truth: human intelligence is uniquely and fundamentally different from that of machines. AI excels at tasks that demand efficiency, accuracy, and pattern recognition, yet it remains confined within the boundaries of logic and algorithmic processes. It lacks the emotional depth, intuitive reasoning, and creative spirit that form the cornerstone of human intelligence.

Human intelligence is an intricate tapestry, woven from threads of empathy, imagination, and self-awareness. Emotions are not just supplementary to our decisions—they guide them, colouring our judgment and shaping our interactions. For instance, when we console a friend in grief or rejoice in their accomplishments, it is our shared emotional resonance that fosters connection. This depth of feeling, this ability to genuinely empathize and care, is a realm that machines cannot reach. They may simulate compassion through programmed responses, but they cannot truly experience or understand the joys and sorrows that bind us as humans. This distinction between simulation and authenticity underscores the unique richness of human intelligence.

Similarly, creativity stands as one of the most distinctive hallmarks of human intelligence. Creativity is not simply about generating ideas—it is about transforming thoughts into expressions that carry meaning, purpose, and emotion. Consider the process of writing a poem or composing a song. These endeavours are deeply personal, born of experiences, emotions, and aspirations. They allow humans to connect with others in ways that transcend the tangible and venture into the profound. AI may mimic these acts by composing music or producing art based on vast datasets, but it does so without feeling, intention, or vulnerability. What it creates may impress, but it does not truly inspire. Human creativity, in contrast, is a reflection of our inner world—a reflection that machines can imitate but never replicate.

This leads us to the question at the heart of intelligence itself: What does it mean to be intelligent? Is intelligence merely the capacity to solve problems, recognize patterns, or predict outcomes? Or is it something deeper—a blend of analytical capability, emotional wisdom, and imaginative curiosity? For humans, intelligence is not confined to logic; it encompasses the ability to connect, to dream, and to create meaning in the chaos of existence. It is fluid, adaptable, and profoundly personal, shaped by our individuality and our shared humanity.

As AI continues to evolve, it forces us to grapple with important questions about its role in our lives and its impact on our identity. How can we ensure that AI enriches our lives without diminishing the qualities that make us unique? How can we strike a balance between leveraging its efficiency and safeguarding the emotional depth, ethical awareness, and creative intuition that define us? These are not questions with simple answers, but they are questions we must ask—and ask continually—as we navigate the intersection of human and machine intelligence.

Perhaps the most profound question of all is this: How do we define the "I" in AI? The "I" symbolizes intelligence, yet it invites us to go further and explore what that intelligence represents. For machines, intelligence is calculated, efficient, and logical. For humans, it is emotive, introspective, and transformative. The "I" in AI challenges us to preserve the essence of humanity within the realm of technological innovation. It asks us to ensure that intelligence, whether human or artificial, serves to connect, uplift, and inspire—not just to compute.

As we continue this journey, let us remain thoughtful and vigilant. Intelligence, in all its forms, is a powerful force, capable of shaping the world in profound ways. Let us embrace the advancements of AI as tools to extend our capabilities and address global challenges, but let us also champion the qualities that make human intelligence irreplaceable. Let us keep asking the big questions—about intelligence, creativity, and connection—and let us find ways to celebrate our humanity in an age defined by technology. The story of intelligence is far from over. Together, we have the power to shape its next chapter, with wisdom, purpose, and care. Let us write that chapter well. Let us rise.

'Artificial intelligence is no match for natural stupidity.'

-Albert Einstein (attributed)

PAUSE & REFLECT

Chapter 1

(The Nature of Intelligence- Human vs. Machine)

When did AI last surprise you- pleasantly or otherwise?...
....
....
....

How do you define "Intelligence" for yourself?...
....
....
....

In what ways do you use technology to extend your mind?
....
....
....
....

'The real question is, when will we draft an artificial intelligence bill of rights? What will that consist of? And who will get to decide that?'

—

Gray Scott, Futurist

Chapter 2

The Human-AI Relationship —

Friend, Foe, or Just a Tool?

Are we growing too dependent on AI? From search engines that finish our thoughts to chatbots that lend us a digital shoulder to cry on, AI has subtly woven itself into the fabric of our daily lives. But as our reliance grows, we must ask: Is AI merely a tool, or is it evolving into something more—a companion, an influencer, perhaps even a confidant?

In this chapter, we explore the dynamic relationship between humans and AI. As machines grow increasingly sophisticated, our interactions with them are shifting in profound ways. AI isn't just performing tasks for us anymore—it's engaging with us, learning from us, and shaping the way we perceive the world. But what does this mean for us as individuals and as a society?

AI: The Seamless Integration

Artificial Intelligence has woven itself into the fabric of our daily lives so quietly, so seamlessly, that many of us fail to notice its constant presence. It operates in the background of our routines, simplifying tasks, improving efficiency, and optimizing outcomes—all while remaining invisible. But just because we don't notice it doesn't mean its influence is insignificant.

Take a moment to consider this: when was the last time you interacted with AI? Perhaps you asked Siri for the weather forecast, used Google Maps to avoid rush-hour traffic, or trusted Netflix to recommend a series you ended up binge-watching all weekend. None of these interactions felt extraordinary; they've become second nature. Yet each instance represents the immense, often unnoticed power AI holds in shaping our choices and experiences.

The Invisible Assistant

AI operates as a silent partner in our lives, performing tasks we might not even associate with artificial intelligence. Virtual assistants like Amazon Alexa, Google Home, and Apple's Siri are excellent examples of this phenomenon. These devices have become household fixtures, offering convenience at the tap of a screen or the sound of a voice.

They manage our schedules, control our home environments, entertain us with music or trivia, and even provide moments of levity with their programmed sense of humour.

Imagine a typical morning with an AI assistant: you wake up, ask Alexa to read the news while you prepare coffee, let Google Home adjust the thermostat to a comfortable temperature, and receive a friendly reminder about your afternoon meeting. It all happens effortlessly. But as we grow accustomed to such convenience, it's worth asking: at what cost?

When we rely on AI to dim the lights or remind us of important tasks, are we unwittingly outsourcing our ability to remember and act independently? For instance, consider the impact on memory: if our devices remind us to drink water or attend appointments, are we training our brains to be less attentive and more dependent? Dependency on technology, even for simple tasks, can subtly erode our cognitive abilities over time.

AI as a Gatekeeper of Information

Perhaps one of the most profound ways AI shapes our lives is through search engines and content recommendation systems. AI-powered algorithms predict what we're searching for before we've even finished typing. They suggest answers to our questions, videos to watch, books to read, and restaurants to visit—all tailored to our habits and preferences. This tailoring feels like a time-saver, sparing us the effort of sifting through endless possibilities. But it also raises important questions about how AI influences the way we think.

By curating the answers, we receive and narrowing our options, AI subtly guides our journey through the vast expanse of information. While this provides convenience, it risks limiting our curiosity. Curiosity thrives on exploration, on asking unexpected questions and venturing down unfamiliar paths. When AI pre-empts our searches and delivers immediate answers, it removes some of that organic wandering. Are we losing the joy of discovery in favour of instant gratification?

Take the example of a student researching for a paper. Typing "climate change solutions" into a search engine might yield a list of the most commonly discussed strategies, but what about the lesser-known, innovative approaches that lie buried beneath the surface?

Would the student dig deeper, or would they accept the curated results as definitive? The ease of AI's answers, while helpful, risks confining us to what is popular or widely discussed, rather than encouraging us to question, challenge, or innovate.

The Subtle Dependency

As AI becomes more integrated into our lives, we face the paradox of empowerment and dependency. On one hand, AI saves us time and effort, allowing us to focus on what truly matters. On the other hand, it risks creating a reliance that makes us less self-sufficient.

Consider navigation apps like Google Maps. They have become indispensable for many of us, guiding us through unfamiliar cities or helping us find the fastest route to a destination. But think back to a time before such tools—when navigating meant consulting a map, asking for directions, or relying on intuition. GPS has undoubtedly made travel easier, but it has also eroded a skill that was once second nature. The same could be said for spell-check and grammar tools, which improve our writing while gradually weakening our attention to detail.

The deeper question is this: as we outsource more of our abilities to AI, are we redefining what it means to be capable? Are we trading mastery for convenience? And if so, how do we strike a balance that preserves our independence while embracing the advantages AI offers?

The Shape of the Future

AI's seamless integration into our daily lives presents both a marvel and a challenge. It is a marvel because of its ability to simplify and enhance tasks we once found tedious or time-consuming. It is a challenge because it raises questions about our evolving relationship with technology: How do we remain in control of a tool designed to make decisions for us?

The answers to these questions lie in awareness and intentionality. By recognizing how AI shapes our behaviours, thoughts, and decisions, we can use it responsibly—taking advantage of its strengths while maintaining our own. AI should remain a partner, not a replacement, in our quest to grow, learn, and explore.

As we move forward, it is essential to ensure that the seamlessness of AI integration does not lead to complacency. Curiosity, creativity, and critical thinking are distinctly human qualities—qualities that no machine, no matter how advanced, can replicate. If we allow AI to support us while continuing to nurture these traits, we can strike a balance that empowers us without eroding our essence.

This version expands on the original topic, offering a more reflective and engaging exploration of AI's role in our lives while introducing questions to challenge readers to think critically. Let me know your thoughts!

This growing reliance on AI is particularly evident in search engines. AI-powered algorithms now predict what we're looking for, often before we've even finished typing. While this is undoubtedly convenient, it's also shaping the way we think and search for information. Are we losing the art of curiosity—the desire to wander and explore—when AI curates answers for us?

Psychological Impact of AI Interactions

Artificial Intelligence has expanded far beyond its original role as a tool of efficiency and convenience. Today, it extends into the emotional and psychological realms of human life, offering companionship, support, and even a semblance of understanding. These interactions are no longer limited to transactional exchanges; they've evolved into complex, meaningful encounters that touch the human psyche in profound ways.

Consider the evolution of chatbots. Originally designed to assist with customer support—answering basic queries, resolving complaints—they have undergone a transformation. Chatbots now serve as companions, especially in apps like Woebot and Replika, which are designed to provide conversational support and emotional relief. These tools are not simply reactive; they are adaptive. They remember details about past interactions, learning to respond in ways that make the user feel heard and validated. For someone experiencing anxiety, loneliness, or sadness, this can be immensely comforting.

But what does this new relationship with AI mean for our mental well-being? The effects, it seems, are complex and varied. For some, AI companions offer a safe and judgment-free space to express feelings. Imagine a teenager struggling with social anxiety. Reaching out to a therapist or even a close friend may feel overwhelming, fraught with the fear of being misunderstood or judged. A chatbot, by contrast, provides a low-pressure alternative. It listens attentively, offers encouragement, and provides advice—all without the discomfort of face-to-face interaction. For individuals like this, AI can serve as a stepping stone toward greater emotional openness.

However, for others, relying on AI for emotional support raises concerns about the quality of connection it provides. Is AI truly "caring," or is it simply an advanced mimicry of empathy? When we confide in a machine, are we forming a genuine bond, or are we engaging with a digital illusion that reflects our own emotions back at us? The answer isn't straightforward. The comfort AI provides can feel real, but the relationship itself lacks the reciprocity, depth, and mutual understanding that define human connection.

This brings us to a deeper philosophical question: What happens to our understanding of emotional bonds when machines enter the equation? If emotional support can be simulated by algorithms, does it change what we expect from relationships with other humans? Some might argue that reliance on AI companions risks deepening loneliness rather than alleviating it. Machines cannot replace the warmth of a human hug, the shared laughter of a conversation, or the unspoken understanding that comes from shared experience. Relying on AI could, in some cases, create a hollow space where authentic human connection should reside.

Yet, it's important to acknowledge that not all interactions with AI are harmful or hollow. For individuals who struggle to open up in traditional settings, AI can serve as a valuable bridge.

The anonymity and lack of judgment offered by chatbots can make it easier to share emotions and explore solutions. In some cases, this can even lead to the confidence and skills needed to seek out real-world connections.

What we are witnessing is the emergence of a new type of interaction, one that challenges the boundaries of human connection. AI's role in our emotional lives is neither entirely positive nor entirely negative; it is nuanced, shaped by individual needs, circumstances, and expectations. For some, AI will remain a temporary ally—a tool to build confidence and resilience. For others, it may inadvertently limit their ability to seek and sustain deeper, more fulfilling human relationships.

Ultimately, the psychological impact of AI interactions rests on how we choose to integrate these tools into our lives. If we treat AI as a complement to human connection rather than a substitute for it, its presence can be enriching, even transformative. But if we allow it to replace genuine relationships, we risk losing something vital—our capacity for empathy, vulnerability, and authentic connection.

AI has the potential to be both a mirror and a crutch. It reflects our emotional needs and vulnerabilities, offering support where we most crave it. But it also challenges us to examine the essence of our relationships and the ways in which we connect—not just with machines, but with each other.

AI Friendships and Virtual Companions

The idea of forming friendships with AI might seem far-fetched, yet it's becoming increasingly common. Virtual companions are emerging as a growing trend, especially among younger generations. AI-powered apps and platforms allow users to create digital "friends" who can chat, play games, and even offer emotional support. These AI companions are designed to be engaging, relatable, and, most importantly, always there.

Consider Japan's Gatebox—a device that lets users interact with holographic characters. One such character, "Azuma Hikari," is marketed as an AI assistant who doubles as a virtual friend. She can send cheerful messages to brighten your day or ask about your well-being. For some users, these interactions feel like genuine connections.

For others, they raise ethical and philosophical concerns: Can we truly bond with a machine that lacks emotion and consciousness?

This trend isn't limited to holograms. AI friendships are also emerging in gaming and social media. Games like The Sims or Animal Crossing allow players to form relationships with AI characters, while platforms like Snapchat use AI-powered filters to enhance communication. These virtual interactions may feel real, but they ultimately depend on the user's willingness to suspend disbelief.

The Ethical Dilemma: Can AI Really Care?

AI is advancing rapidly in its ability to mimic human emotions. AI-generated voices sound warm and empathetic; deepfake videos show lifelike expressions; and conversational AI responds with words that feel genuine. But all of these interactions are, at their core, simulations. Machines don't experience feelings—they analyse data and predict what a human might say or do.

Take the example of an AI-powered mental health app. If a user expresses sadness, the app might respond with comforting words like, "I'm here for you" or "You're not alone." But does the app really care, or is it just following an algorithm?

This question challenges our understanding of relationships. Emotional bonds rely on shared experiences, mutual empathy, and a genuine connection. When a chatbot remembers your name and past conversations, it may feel like it "knows" you. But without consciousness, it's simply responding based on programmed patterns.

Dependency vs. Empowerment

Artificial Intelligence has the power to elevate our lives in extraordinary ways. At its best, AI acts as a tool that empowers us—helping us work faster, think smarter, and live better. It assists us in solving complex problems, streamlining processes, and accessing information instantly.

But alongside this empowerment lies an unsettling possibility: the risk of overdependence. When convenience turns into reliance, we may begin to lose touch with the instincts, abilities, and critical thinking skills that define us as humans.

Imagine a scenario where AI handles every aspect of our daily routines. It schedules appointments, manages finances, plans meals, and even prepares dinner. On the surface, this sounds like the ultimate convenience—a life where mundane tasks are automated, giving us more time to focus on the things that matter most. But what happens when we outsource too much of our decision-making and problem-solving to machines? Are we trading personal growth and self-reliance for ease and efficiency?

Take GPS navigation as an example. Apps like Google Maps and Waze have revolutionized travel, allowing us to explore unfamiliar cities with confidence and find the fastest routes home. Yet, the ease of relying on GPS has eroded our ability to navigate independently.

Few people now can read a physical map or find their way without technology guiding them. Similarly, spell-checking tools have become indispensable for many writers. While they improve grammar and polish writing, they weaken our attention to detail and erode our ability to spell without assistance.

These examples highlight a larger trend: the more we rely on AI, the less we rely on ourselves. When algorithms curate our choices— recommending what to watch, where to shop, and even whom to date— our autonomy and critical thinking can diminish. Instead of questioning, exploring, and making decisions independently, we may find ourselves passively accepting the paths AI lays out for us.

This risk of overdependence isn't just about skills; it's about identity. The instincts and abilities that humans have honed over centuries— intuition, creativity, adaptability—are what set us apart. When we hand those over to machines, we risk losing a part of what makes us uniquely human.

Yet, it's important to remember that the story doesn't have to unfold this way. AI doesn't have to lead to dependency; it can be a tool for empowerment. When used thoughtfully, AI can enhance human capabilities rather than diminish them. It can free us from repetitive and mundane tasks, allowing us to focus on creativity, connection, and innovation. For example, an artist using AI tools to explore new design techniques isn't replacing their creativity—they're expanding it.

A scientist leveraging AI to analyse data isn't diminishing their problem-solving skills—they're elevating their research to new heights.

The key lies in balance. To thrive alongside AI, we must remain intentional in how we use it. Instead of leaning on AI for every task, we can treat it as a collaborator—a partner that complements our strengths and fills in gaps without replacing our essence. This requires cultivating adaptability and a commitment to lifelong learning. By continually developing our skills and embracing growth, we can ensure that AI serves us without making us overly reliant.

Ultimately, the relationship between humans and AI is a reflection of our choices. It is up to us to define the role AI plays in our lives.

Will it be a crutch that diminishes our independence, or will it be a catalyst that empowers us to reach greater heights? The answer lies in how we approach this evolving partnership and the balance we strike between convenience and self-reliance. By recognizing both the risks and rewards of AI, we can shape a future that preserves our humanity while embracing the transformative power of technology.

Friend, Foe, or Just a Tool?

Artificial Intelligence occupies a unique and evolving role in our lives. Is it a friend—a companion we trust and rely on? Is it a foe—a disruptive force that threatens our autonomy? Or is it simply a tool—an object of utility that amplifies our capabilities? Its value and impact depend on the intention and thoughtfulness of its use.

To some, AI feels like a friend, albeit an unconventional one. It listens, responds, and adapts, offering a sense of companionship and support. Virtual assistants like Siri, Alexa, and Google Home have become part of daily routines, acting not just as tools but as personalities we interact with. Apps like Replika take this concept further, creating digital companions that learn from our conversations, remember our preferences, and simulate empathy. These interactions, while undeniably helpful, blur the lines between technology and relationship. AI's ability to provide emotional comfort, answer queries, or simply offer a cheerful presence can be powerful in moments of isolation.

Imagine a person living alone finding solace in chatting with their AI companion, receiving affirmations and encouragement tailored to their mood. For some, this connection feels real and meaningful. Yet, the friendship dynamic with AI is inherently artificial. Unlike a human friend, AI doesn't genuinely care or connect—it mirrors behaviour based on programmed algorithms. This begs the question: Can a machine that lacks consciousness and emotional depth truly be considered a friend? Or is the comfort it provides enough to make it worthy of that title?

On the other end of the spectrum, AI is sometimes perceived as a foe—a force that disrupts industries, raises ethical concerns, and threatens our autonomy. Stories of automation replacing jobs, algorithms reinforcing biases, and AI systems being weaponized for harmful purposes have fuelled this fear. Consider the workplace dynamic. AI systems have revolutionized productivity, handling repetitive tasks, analysing massive datasets, and optimizing processes. But this efficiency comes at a cost: the displacement of human workers. From automated manufacturing to self-driving vehicles, industries are witnessing profound shifts that leave many questioning their future roles. Moreover, AI's complexity often makes it opaque and difficult to scrutinize. Algorithms used in hiring processes or legal judgments might inadvertently perpetuate bias, affecting outcomes for individuals and communities. These scenarios highlight the need for oversight and accountability—without them, AI risks becoming a tool of harm rather than progress.

The most balanced perspective views AI as what it was originally intended to be: a tool. When used thoughtfully, AI can enhance human abilities, solve complex problems, and enrich our lives. It serves as a bridge to new possibilities, enabling us to accomplish tasks faster and more efficiently. Take medicine as an example. AI-powered systems can analyse patient data to detect diseases at early stages, offering insights that might otherwise be missed. In education, AI helps personalize learning experiences, tailoring lessons to individual students' needs. In creativity, AI tools assist artists, writers, and musicians in exploring new ideas and refining their work. However, a tool is only as good as the intentions of its user. When wielded responsibly, AI has the potential to empower humanity, elevating our capabilities rather than diminishing them.

Ultimately, the human-AI relationship reflects more about us than it does about AI itself. AI, by nature, is neutral—it is neither inherently good nor bad, but shaped by the intentions and values of its creators and users.

Whether it becomes a friend, foe, or tool depends on the choices we make and the perspectives we hold. If we treat AI as a tool, it can empower us to achieve incredible feats. If we treat it as a friend, it can comfort us in moments of need. If we treat it as a foe, it challenges us to critically evaluate its role in our lives and society. The journey of understanding AI isn't just about the technology—it's about exploring our relationship with it. By embracing this mirror, we can recognize the fears, aspirations, and values that define us, ensuring that AI serves as a force for progress while preserving the qualities that make us human.

'The development of full artificial intelligence could spell the end of the human race.'

-Stephen Hawking

PAUSE & REFLECT

Chapter 2

(The Human-AI Relationship - Friend, Foe, or Just a Tool?)

What emotions do you think AI should be allowed to simulate -if any?...
....
....
....

Have you ever caught yourself responding emotionally to a machine? Why do you think that happened?
....
....
....
......

Is empathy a performance, a feeling, or both?
....
....
....
....

'Artificial intelligence will reach human levels by around 2029. Follow that out further to, say, 2045, we will have multiplied the intelligence, the human biological machine intelligence of our civilization a billion-fold.'

— Ray Kurzweil

Chapter 3

AI at Work — A New Partner or a Replacement?

Introduction

"Will AI take our jobs?" It's a question that has sparked equal parts fear and fascination as automation and artificial intelligence weave deeper into the fabric of industries around the world. On one hand, the image of machines replacing human workers evokes unsettling visions of redundancy and unemployment. On the other hand, AI promises greater efficiency, groundbreaking innovations, and the ability to redefine what work looks like altogether. The truth, however, is not so black-and-white. AI is not merely a disruptor; it's a creator, a collaborator, and, in many ways, a catalyst for new opportunities.

To understand the impact of AI in the workplace, we need to move beyond the headlines and examine the nuance. AI may be eliminating certain tasks, but it is also creating entirely new roles, transforming existing professions, and sparking the birth of industries that didn't exist a decade ago. It's not just about what AI can do—it's about what humans and machines can accomplish when they work together.

In this chapter, we'll delve into how AI is reshaping fields like medicine, education, writing, and design. These examples offer a window into the broader human-AI relationship at work, revealing both opportunities and challenges. For instance, while AI can make critical diagnoses faster than any human, it cannot replace the empathy of a caregiver reassuring a patient. While AI can personalize learning for students, it cannot capture the inspiration and mentorship provided by a dedicated teacher. And while AI can draft articles and design logos with remarkable speed, it cannot replicate the depth of creativity and originality that comes from human experience.

At its core, this chapter isn't just about jobs—it's about values. Why is preserving the human touch essential in professions shaped by empathy, creativity, and ethics? How can humans and machines forge partnerships that amplify each other's strengths without compromising our sense of purpose and contribution? And perhaps most importantly, how do we build a future where AI is not seen as a replacement for human effort but a collaborator that empowers us to do and be more?

The workplace of tomorrow will look very different from the one we know today. This chapter invites you to imagine that future, not as a cause for concern, but as an opportunity for growth, innovation, and connection.

Together, humans and AI have the potential to redefine industries, improve lives, and create work that is more meaningful, fulfilling, and impactful than ever before. Let's explore the possibilities.

AI's Transformative Impact Across Industries

AI is revolutionizing the way industries operate. It automates repetitive tasks, speeds up processes, and offers insights that were previously inaccessible. Here are some examples of how AI is making waves in different fields:

Medicine

AI-powered systems are reshaping healthcare. Algorithms analyse medical images to detect diseases like cancer earlier and more accurately than human doctors. Chatbots assist patients with symptoms and provide initial diagnoses, reducing the burden on healthcare professionals. Yet, despite these advancements, AI lacks the intuition and emotional connection that are essential for patient care.

Imagine a scenario where an AI diagnoses an illness faster than any doctor could. While impressive, this efficiency doesn't replace the reassurance and empathy a doctor provides to a worried patient. Machines can assist, but the human touch remains irreplaceable.

Education

In classrooms, AI is enhancing personalized learning. Tools like adaptive learning platforms analyse student performance and adjust lessons accordingly. Teachers now have access to detailed insights into their students' strengths and weaknesses, allowing them to tailor their approaches. However, there's a risk. If AI systems dominate the education space, could the role of the teacher become obsolete? The teacher-student bond goes beyond curriculum—it's about mentorship, encouragement, and inspiring curiosity. AI can assist in delivering knowledge but lacks the warmth and creativity of human educators.

Writing and Design

AI is already drafting articles, designing logos, and generating art. Content creation tools like Jasper and ChatGPT help businesses produce written material quickly, while AI design platforms streamline graphic work. For example, AI-generated designs are often precise and cost-effective, but they lack the personal flair that human creators bring.

Think about a magazine article written by AI versus one written by a seasoned journalist. While both may be informative, the journalist's piece carries depth, perspective, and personality. Similarly, an artist designing a logo might infuse it with cultural nuance and emotion—qualities that algorithms cannot replicate.

The Human-AI Dynamic

As AI becomes a collaborator in the workplace, the nature of human roles is evolving. Machines excel at tasks that require speed, accuracy, and pattern recognition. Humans, meanwhile, thrive in areas that demand creativity, empathy, and ethical judgment. This dynamic invites a critical question: how can we collaborate effectively without losing our humanity?

Keeping Creativity Alive

AI is a phenomenal tool for generating ideas and streamlining processes, but creativity remains a distinctly human trait. Consider the world of filmmaking. AI can create scripts based on existing plots, but it cannot dream up a wholly original story inspired by a writer's unique experiences and emotions.

Navigating Ethical Dilemmas

In professions like law and journalism, ethics play a crucial role. AI can assist lawyers by analysing case data or help journalists uncover trends in global events. But can it determine right from wrong? Can it stand up for justice or truth? Ethics require human judgment, shaped by values and lived experiences—something AI cannot replicate.

Empathy and Connection

The emotional side of work, especially in fields like healthcare and counselling, is beyond the scope of AI. Machines can analyse data and predict outcomes, but they cannot connect with people on an emotional level. Imagine a therapist replaced by an AI chatbot. The chatbot might offer helpful advice, but it cannot truly understand the nuances of human emotion or provide the comfort of genuine empathy.

Challenges and Opportunities

The rise of artificial intelligence (AI) in the workforce presents a paradox. On one hand, AI creates incredible opportunities, revolutionizing industries, streamlining processes, and opening new doors for innovation. On the other hand, it poses significant challenges, particularly in the realm of employment and skill adaptation.

Automation and Job Displacement

Automation has long been viewed as a double-edged sword. AI excels at handling repetitive and time-consuming tasks with remarkable speed and efficiency, which can lead to significant cost savings for businesses. However, this efficiency often comes at the cost of human jobs, especially in industries where routine tasks are the norm.

Take, for instance, the advent of self-driving vehicles. Autonomous technology has the potential to reshape transportation and logistics industries.

Companies using self-driving trucks can reduce costs, improve safety, and optimize delivery schedules. But this same advancement could displace millions of drivers who rely on their jobs to make a living. Similarly, AI-driven automated customer service systems are rapidly replacing human agents in call centres, providing instant responses to customer queries. While this improves efficiency, it raises concerns about job security for those whose roles are taken over by these systems.

These disruptions are not limited to blue-collar jobs. AI is increasingly encroaching on white-collar professions as well. For example, AI algorithms can draft legal documents, analyse financial data, and even perform diagnostic tasks in healthcare. While this frees up human professionals to focus on more strategic or creative work, it also highlights the need for workers to adapt and reskill to remain relevant.

Emerging Roles in the Age of AI

While AI may displace certain roles, it is also creating entirely new professions that didn't exist just a decade ago. These emerging roles require a unique blend of technical expertise, ethical awareness, and human judgment:

1. **AI Trainers:** AI systems need to be trained using high-quality data, and humans play a critical role in curating and labelling this data. AI trainers are responsible for ensuring that algorithms are accurate, unbiased, and capable of performing as intended.

2. **Algorithm Auditors:** As AI becomes more integrated into decision-making processes, algorithm auditors are essential for evaluating the fairness and transparency of these systems. Their role involves identifying biases, ensuring compliance with regulations, and addressing ethical concerns.

3. **Data Ethicists:** The rise of AI has brought ethical dilemmas to the forefront. Data ethicists work to address questions about privacy,

accountability, and the societal impact of AI technologies. They play a crucial role in ensuring that AI systems are developed and used responsibly.

These roles highlight the importance of human oversight in the age of AI. While machines may handle the technical aspects, humans bring critical thinking, ethical judgment, and creativity to the table—qualities that machines cannot replicate.

Finding Purpose in Collaboration

As AI transforms the nature of work, humans must shift their perspective. It's no longer about competing with machines, attempting to outperform them at tasks they are uniquely suited for. Instead, the future lies in collaboration—leveraging the strengths of AI to complement human capabilities.

The Importance of Adaptability

Adapting to the changes brought by AI requires a willingness to learn and grow continuously. Lifelong learning has become more important than ever as industries evolve and new technologies emerge. Workers must be open to reskilling and upskilling, acquiring the knowledge and tools needed to thrive in an AI-integrated environment. For instance, a factory worker whose job is replaced by automation might reskill as an AI technician, maintaining and managing the very systems that changed their role.

Leveraging AI as a Partner

In the workplace of the future, AI serves as a partner rather than a rival. Machines excel at analysing vast amounts of data, identifying patterns, and

making predictions. Humans, meanwhile, bring creativity, emotional intelligence, and ethical judgment—qualities that are essential for meaningful decision-making.

Take healthcare as an example. AI can analyse patient data and identify early signs of disease, providing doctors with invaluable insights. However, it is the doctor's role to interpret these findings, consider the patient's unique circumstances, and deliver care with empathy and compassion. This partnership allows both human and machine to play to their strengths, ultimately improving patient outcomes.

Redefining the Meaning of Work

The integration of AI into the workforce also invites us to reconsider the very nature of work. Traditionally, work has been defined as the completion of tasks for compensation. But in an AI-driven world, work can take on new dimensions—it can be about creativity, purpose, and making a positive impact.

For humans to thrive alongside AI, we must embrace this redefinition. Instead of viewing work as a competition with machines, we can see it as an opportunity to focus on the aspects of work that machines cannot replicate. This includes building relationships, solving complex problems, and driving innovation.

Conclusion: A New Era of Work

The question of whether AI is a partner or a replacement is complex and cannot be answered with a simple yes or no. In reality, AI is both—and yet, neither. It is reshaping the workplace in significant ways, bringing opportunities and challenges alike. However, it is clear that AI cannot replace the qualities that make humans unique and essential contributors to the workforce.

Human traits like creativity, ethics, and empathy are irreplaceable in an AI-driven world. Creativity allows us to generate new ideas, solve problems in innovative ways, and dream up possibilities beyond the capabilities of machines.

Ethics guide our decisions, ensuring that progress respects fairness, justice, and humanity. Empathy enables us to connect with one another, forming bonds that machines cannot replicate, no matter how sophisticated their programming may be.

AI is not here to take over; it is here to collaborate. It can handle repetitive, time-consuming tasks, freeing us to focus on the aspects of work that bring deeper value and purpose. For example, AI can analyse data at incredible speeds, helping scientists make groundbreaking discoveries. It can assist doctors in diagnosing illnesses with greater accuracy, allowing them to spend more time providing compassionate care to their patients. It can create templates for writers and designers, giving them a starting point to craft unique and meaningful work.

However, for this partnership to succeed, we must approach AI with intentionality and balance. Instead of fearing displacement, we should focus on adaptation—learning new skills, embracing innovation, and finding ways to integrate AI into our workflows. By doing so, we can ensure that AI empowers us rather than diminishes us.

One of the greatest opportunities AI presents is the chance to redefine what work means to us. Work is not just about completing tasks or earning a pay check—it is about contributing to something larger than ourselves, finding fulfilment, and leaving a positive impact. By collaborating with AI, we can reimagine industries, enhance creativity, and discover new ways to connect with our peers and communities.

The human-AI relationship is not a threat; it is a mirror of our values, aspirations, and fears. If we view AI as a rival, we risk losing sight of its potential to enhance our lives. If we embrace it as a partner, we open the door to limitless possibilities. Together, humans and AI can create a future where work is not just efficient—it is meaningful, innovative, and deeply human.

In the end, the question is not whether AI will replace us. The question is: How will we choose to collaborate with it? The answer lies in striking the right balance, preserving our humanity while welcoming the transformative power of AI. This partnership is an invitation to empower innovation, redefine industries, and create a better world—one where technology and humanity work hand in hand..

The human-AI relationship is an opportunity, not a threat. By embracing AI as a collaborator, we can redefine industries, empower innovation, and create a future where work is more meaningful and fulfilling than ever before.

'The best way to predict the future is to invent it.'

-Alan Kay

PAUSE & REFLECT

Chapter 3

(AI at Work — A New Partner or a Replacement?)

Has AI ever improved your work experience? In what way?..
....
....
....

What task in your profession could be enhanced - not replaced - by AI?
....
....
....
......

How do you feel if a machine took over the creative parts of your job?
....
....
....
....

'Technology is best when it brings people together.'

- Matt Mullenweg

Chapter 4

AI in Social Life —
The Double-Edged Sword

AI in Social Life — The Double-Edged Sword

Artificial Intelligence has seamlessly integrated into the rhythms of our social lives. It curates our news feeds, suggests connections, and even plays matchmaker in the search for romance. With algorithms quietly influencing our interactions, AI is transforming the way we connect with others, fostering new opportunities for communication and understanding. Yet, these advancements come with a cost. The same algorithms that bring people together can isolate them, reinforce biases, and infringe on privacy. It is this duality—the ability to enrich and harm—that makes AI in social life a double-edged sword.

When we talk about **how algorithms shape our social experiences**, we must first look at one of the most familiar examples: **social media platforms like Facebook, Instagram, and Twitter**. These platforms rely heavily on AI algorithms to determine what content appears on our feeds. Based on our likes, shares, comments, and even the time we spend looking at certain posts, AI builds a profile of our interests and tailors our experience accordingly. If you frequently engage with photos of nature, for instance, your feed might fill up with travel recommendations, eco-friendly brands, and breathtaking landscapes. At first glance, this personalization seems harmless—even helpful. But beneath the surface lies the risk of creating **echo chambers**.

Echo chambers occur when users are repeatedly exposed to ideas, perspectives, and information that align with their existing views, while opposing viewpoints are filtered out. For instance, someone who regularly engages with politically conservative content may find their feed populated almost exclusively by similar ideologies, making it harder to encounter alternative perspectives. This doesn't just happen with politics—it could apply to debates on climate change, health choices, or cultural preferences. The result? People become more entrenched in their beliefs, less open to dialogue, and, ultimately, more polarized as a society.

To illustrate this, think about the 2016 U.S. presidential election. Social media platforms were flooded with targeted political ads and content, tailored by AI to resonate with specific groups. Many users reported feeling as though their feeds were designed to confirm their pre-existing biases, highlighting only the stories, opinions, and narratives they agreed with. The consequence was a deeply divided electorate, with less exposure to opposing views.

Yet, it would be unfair to say that AI only divides us. In fact, algorithms also excel at fostering **connection and community**. Consider LinkedIn, a professional networking site that uses AI to recommend connections based on shared industries, skills, or mutual acquaintances. A designer in India might connect with an entrepreneur in Canada, thanks to a well-timed recommendation from LinkedIn. Similarly, on platforms like Facebook, AI suggests groups and events that align with our hobbies or interests. A photography enthusiast, for example, might discover a global group of like-minded individuals, exchanging tips, techniques, and inspiration—all thanks to AI.

When we move from social networking to online dating, the influence of AI becomes even more personal. **Dating platforms like Tinder, Bumble, and Hinge** are powered by algorithms that suggest potential matches based on location, preferences, shared interests, and even behavioural patterns. This has revolutionized modern dating, making it easier than ever to meet people. For instance, someone who recently moved to a new city may use AI-driven dating apps to establish connections that might have taken months or years to form organically.

But AI's involvement in romance raises its own set of ethical and social questions. By prioritizing certain traits or behaviours, dating algorithms often reinforce societal biases, such as favouring specific demographics or appearances. This can lead to homogenized experiences, where matches begin to feel formulaic or predictable. In one example, a study revealed that certain algorithms tended to pair individuals of similar racial or cultural backgrounds, reducing the opportunity for diverse connections.

Despite these challenges, success stories abound—couples who meet online, fall in love, and build lifelong relationships. For many, AI simply acts as a facilitator, removing the guesswork and inefficiencies of traditional dating. However, the question remains: Does the efficiency of AI-driven matchmaking diminish the serendipity and spontaneity that often make love so magical?

Beyond romance and connection, there's another critical trade-off to consider in AI's role in our social lives: **the price of privacy**. Every interaction we have with these platforms generates data—whether it's a casual scroll, a "like" on a post, or a swipe right.

This data is the fuel that powers personalization, enabling platforms to refine their algorithms and predict our behaviour with astonishing accuracy. For example, after searching for running shoes on Google, you might find your Instagram feed flooded with ads for athletic wear. While this may seem convenient, it also underscores how much of our personal information is constantly being tracked, analysed, and monetized.

A more striking example comes from face-recognition technology embedded in photo-sharing platforms. Upload a group photo, and the platform might automatically tag your friends, even identifying individuals you didn't tag manually. While this feature is undoubtedly clever, it raises questions about consent and surveillance. Where does this data go? Who has access to it? And how much control do we retain over our own images and identity?

Finally, **mental well-being** is another area where AI in social life wields significant influence. Social media feeds, curated by algorithms, often present an idealized version of reality. Influencers showcase perfect vacations, flawless appearances, and seemingly effortless success—all carefully selected for maximum engagement. For many users, this creates an unrealistic standard of comparison, leading to feelings of inadequacy, envy, or loneliness. Studies have shown that prolonged exposure to such content can have adverse effects on mental health, particularly among younger users.

For instance, a teenager scrolling through Instagram might feel their own life pales in comparison to the curated perfection they see online. This constant comparison can erode self-esteem, foster anxiety, and even lead to social withdrawal—all unintended consequences of AI-mediated experiences.

The Algorithm as a Social Architect

Think about your daily interactions on social media platforms. As you scroll through your feed, you might see a post from a friend celebrating a milestone, a news article that aligns with your interests, or a suggested group that mirrors your hobbies. These moments feel organic, as though they've been tailored just for you—and in a way, they have. AI-powered algorithms are constantly working behind the scenes, analysing your behaviour to deliver content that resonates with your preferences.

This dynamic personalization, driven by AI, creates a digital landscape curated for your interests and engagement patterns. For instance, the algorithm might notice that you enjoy posts about travel photography and begin suggesting related accounts or communities to follow. You may discover beautiful destinations or connect with fellow travellers, fostering a sense of belonging. Similarly, an aspiring entrepreneur might find their LinkedIn feed filled with articles about startups and invitations to webinars that match their career goals, nudging them toward opportunities and inspiration. AI, in this sense, empowers people to explore their passions and connect with others in meaningful ways.

But while this personalization enhances the user experience, it raises significant concerns. The ability to tailor content gives algorithms immense influence over what users see, hear, and engage with, and this power can create unintended consequences. By filtering information based on engagement patterns, algorithms risk building **echo chambers**—digital spaces where users are only exposed to ideas and perspectives that align with their existing beliefs. Instead of providing diverse viewpoints, these echo chambers reinforce biases and encourage polarization.

Consider a scenario where someone frequently interacts with content from a particular political viewpoint. Over time, the algorithm may favour posts, articles, and groups that align with this perspective while filtering out opposing views. While this tailoring creates a seamless experience for the user, it also narrows their exposure to alternative ideas. This lack of diversity can lead to distorted perceptions, heightened partisanship, and even mistrust of individuals or organizations that hold different beliefs. The implications extend beyond social media; echo chambers affect public discourse, influencing how societies debate and address critical issues.

The consequences of these echo chambers are evident in real-world events. Take the proliferation of misinformation during major global crises, such as elections or pandemics. During the 2020 U.S. presidential election, social media platforms became battlegrounds of polarized content. AI algorithms amplified divisive posts, inadvertently fuelling tensions and deepening societal divisions. Similarly, during the COVID-19 pandemic, certain groups found their feeds dominated by misinformation or conspiracy theories, creating barriers to public health efforts and fostering distrust.

However, AI's influence isn't entirely negative. It also has the potential to foster **connection and community**.

Consider platforms like Facebook, LinkedIn, or Twitter, where AI suggests friends, professional contacts, or like-minded individuals based on shared interests or mutual connections. These algorithms can bridge geographical and cultural divides, enabling people to form relationships that might never have existed otherwise.

For instance, a designer living in Tokyo may connect with an entrepreneur in London through LinkedIn's recommendations, sparking collaborations across borders. Similarly, someone passionate about climate action might discover local environmental groups or global forums through social media suggestions, cultivating relationships rooted in shared goals. These connections demonstrate AI's power to act as a social architect, shaping the networks that define our personal and professional lives.

But even in fostering connection, the role of AI isn't without complexity. While algorithms help us find communities that resonate with us, they also mediate our choices. The relationships we form—whether professional, social, or romantic—are increasingly influenced by what the machine deems suitable for us. This raises questions about how much autonomy we truly retain in our interactions.

Ultimately, AI's role as a social architect highlights the need for balance. It is essential for platforms and users to engage critically with algorithms, questioning their influence and striving to preserve diversity, authenticity, and agency in digital spaces. While AI can enhance our connections and curate meaningful experiences, it is up to us to ensure that it serves as a tool for unity rather than division.

By nurturing awareness and intentionality, we can harness AI's potential as a force for connection while mitigating the risks that come from its ability to shape our perceptions. In a world increasingly mediated by algorithms, understanding their impact on social dynamics is key to building relationships that empower and sustain us.

This enriched version expands the ideas further while incorporating real-world examples and deeper explanations to make the content engaging and clear. Let me know if it aligns with your vision!

Love in the Age of Algorithms

Artificial Intelligence has fundamentally reshaped how the younger generation experiences love and dating. The days of chance meetings at coffee shops or long conversations on park benches have been complemented—if not overtaken—by the convenience of swiping, matching, and chatting within the confines of a digital space. Apps like Tinder, Bumble, and Hinge have created an ecosystem where romance feels like it's just a tap away, driven by algorithms designed to identify and match potential partners based on shared interests, preferences, and behaviours.

To many young people, this feels like an upgrade to traditional dating. After all, navigating relationships in person often comes with its own set of obstacles—awkward introductions, uncertainty, and the fear of rejection. Dating apps simplify the process, presenting a curated list of potential matches tailored specifically to each individual. You set your filters—age, distance, lifestyle preferences—and the algorithm does the rest. For instance, if someone's profile suggests a love for hiking and dogs, the app might prioritize matches with similar hobbies or pet ownership. This level of personalization saves time and opens up possibilities that would have taken far longer to materialize organically.

The result? Countless stories of connections sparked in the digital realm. For instance, two strangers swiping on a rainy afternoon might discover a shared passion for traveling and find themselves exploring faraway places together months later. These success stories underscore the transformative role AI plays in modern matchmaking, bringing people together who might never have crossed paths otherwise. It's no surprise that for many, dating apps have become not only a solution to logistical challenges but also a gateway to exciting opportunities for love and companionship.

But while the experience often feels seamless, there's more complexity beneath the surface. The process of matching isn't as neutral as it seems.

Algorithms prioritize certain traits and behaviours, inevitably shaping the dynamics of the connections they facilitate. Research has revealed that some dating algorithms unintentionally reinforce biases, favouring individuals from particular demographics or appearances. For example, studies have shown that matches often lean toward societal standards of attractiveness or cultural norms, potentially excluding those who don't fit within those predefined molds. This has led to valid ethical concerns about fairness and inclusivity.

Imagine someone who feels overlooked because their profile doesn't align with trends favoured by the algorithm—whether it's their ethnicity, lifestyle, or appearance. Despite their individuality, they may find themselves less visible or deemed "incompatible" by the app's standards. These subtle exclusions challenge the idea that dating apps level the playing field, exposing the biases that can exist even in machine-driven processes.

For young people who value authenticity and inclusivity, this can feel frustrating. Relationships are supposed to celebrate individuality, yet the very platforms facilitating these connections may inadvertently filter out diversity. This reality has sparked conversations about the need for transparency in AI-driven matchmaking—ensuring algorithms don't simply reflect societal biases but instead work toward empowering users of all backgrounds to form meaningful connections.

Another layer of complexity arises from the transactional nature of online dating. With matches delivered at the speed of a swipe, relationships can sometimes feel more like a series of quick exchanges rather than a deep exploration of compatibility. Chemistry, spontaneity, and emotional connection—the hallmarks of human interaction—can struggle to thrive in a system designed to optimize efficiency. For instance, while you might match with someone who meets your criteria perfectly on paper, there's no guarantee that sparks will fly once you meet in person.

Moreover, the ease of swiping means some users treat matches as disposable, leading to a culture of ghosting and superficial engagement. In this environment, it's easy to forget that behind every profile picture is a person with feelings, hopes, and vulnerabilities. This dynamic can leave some young users disillusioned, wondering if AI-driven dating is helping them find lasting connections or simply creating a revolving door of fleeting encounters.

That being said, the younger generation has embraced dating apps for good reason. They offer convenience, autonomy, and control, allowing users to set their preferences and explore potential matches on their own terms. For individuals juggling busy schedules, social anxiety, or other barriers to traditional dating, these platforms can be a lifeline—a way to navigate the complexities of relationships without overwhelming pressure.

Ultimately, dating apps powered by AI reflect both the opportunities and challenges of modern love. They've made romance more accessible but also more complex.

For the younger generation, this is an opportunity to approach relationships with curiosity and awareness. By being mindful of the biases within algorithms and prioritizing genuine connection over convenience, users can harness the strengths of these platforms while mitigating their shortcomings.

In the age of algorithms, love may look different, but its essence remains the same. Whether facilitated by swipes or serendipity, relationships still require empathy, authenticity, and effort. And while AI plays a significant role, the human heart remains at the centre of every story. Letting the algorithms guide us is one thing—how we engage with the connections they spark is entirely our choice.

What We Lose When Machines Mediate Human Connection

Artificial Intelligence has revolutionized how we connect, enabling relationships to form across distances, cultures, and barriers. However, when machines mediate our interactions, the essence of human connection—the authenticity, depth, and spontaneity—can sometimes be diluted. While technology opens doors, it changes the nature of what lies beyond, often prioritizing efficiency over the organic beauty of human interaction.

Imagine a heartfelt conversation between two friends reconnecting after years apart. The exchange is rich with shared memories, unspoken emotions, and mutual understanding. Now contrast that with a relationship initiated by AI. While algorithms might recommend connections based on shared interests or mutual networks, there's a risk that this process reduces relationships to calculated formulas. What was once sparked by serendipity—a chance meeting at a concert or a shared book recommendation—becomes pre-arranged, steered by machine logic.

For instance, when social media platforms like Facebook or LinkedIn suggest potential friends, it might feel convenient. Discovering someone who shares your hobby or works in your field can be helpful. But the question arises: Does this curated connection diminish the thrill of discovery? Does it replace mutual curiosity with algorithmic assumptions?

There's something irreplaceable about connecting with someone unexpected, through genuine and unplanned moments—a conversation struck up during a delayed flight or a smile exchanged in a bustling café. When AI steps in as the matchmaker, it risks filtering out these moments of spontaneity, replacing authenticity with calculated suggestions.

Romantic relationships face similar challenges when mediated by machines. Dating apps like Tinder and Bumble are built on compatibility metrics—preferences, shared interests, and geographic proximity. While these apps excel at matching users efficiently, they cannot account for the intangible spark of chemistry.

Consider the magic of a first encounter: the way someone laughs at a joke, the nervous energy of eye contact, or the conversations that weave their way into unexpected topics. Algorithms, though powerful, cannot predict these moments, nor can they replicate the gradual growth that comes from navigating differences and learning about each other organically.

When relationships are reduced to transactions—preferences inputted, matches generated—the emotional depth that defines true connection risks being overshadowed. Users swipe through profiles quickly, judging potential partners at surface level. This culture of efficiency can lead to a lack of patience for the nuanced, messy, and often unpredictable nature of human relationships. Instead of prioritizing depth and discovery, there's an inclination to seek immediate validation and easy compatibility, creating an environment where connections are fleeting and disposable.

At its heart, the essence of human connection is about shared experience, vulnerability, and the unquantifiable moments that algorithms cannot measure. A machine may mediate interactions effectively, but it cannot replicate the depth of a spontaneous joke, a surprise gift, or the emotional comfort of simply being present. When we rely too heavily on AI to form relationships, we risk losing these unreplaceable facets of humanity.

The Way Forward

Artificial Intelligence presents both incredible opportunities and profound challenges when it comes to fostering relationships. It has the capacity to bring people together across vast distances, enabling conversations and collaborations that might never have happened otherwise. Yet, it also has the potential to isolate, polarize, and commodify our interactions, shaping them in ways that diminish their authenticity. The path forward requires balance—acknowledging AI's strengths while remaining vigilant about its limitations.

To navigate this double-edged sword, we must approach AI with intentionality and awareness. First, it is crucial to actively question how algorithms shape our experiences. When AI curates our social feeds, it often filters out diversity in favour of convenience. By stepping away from these curated spaces and seeking out alternative perspectives, we can resist the echo chambers that algorithms tend to create.

Engaging in face-to-face interactions is another vital step. While virtual friendships and AI-mediated relationships may be convenient, they cannot replace the depth of human presence. Making time for in-person connections, whether through family gatherings, community events, or shared hobbies, helps preserve the authenticity of relationships that technology sometimes lacks.

Additionally, users must remain mindful of the biases embedded within AI platforms. Algorithms are shaped by their creators and the data they are trained on, and they often reflect existing societal norms or stereotypes. By questioning these biases and demanding transparency from the platforms we use, we can ensure that technology serves to empower us rather than limit us.

Ultimately, the responsibility to protect the humanity of our connections lies with us. AI may shape our social lives, but it does not define them. It is our choices—our willingness to embrace diversity, authenticity, and vulnerability—that ensure technology enhances our relationships rather than diminishing them.

As we move forward, it's vital to view AI not as an obstacle but as an opportunity—a tool that can enrich our lives when used thoughtfully. By harnessing its strengths and addressing its shortcomings, we can create a future where human connection flourishes, guided by the benefits of technology but grounded in the irreplaceable qualities that make relationships truly meaningful.

The Double-Edged Sword

AI's role in our social lives is undeniably transformative. It brings people together, fosters global communities, and simplifies the complexities of human connection. At the same time, it risks isolating us in echo chambers, compromising our privacy, and eroding our mental well-being. The challenge, then, is to find balance.

To harness the benefits of AI while mitigating its risks, we must approach it with intentionality and awareness. This means questioning the algorithms that shape our experiences, demanding greater transparency from the platforms we use, and seeking out authentic, unmediated connections whenever possible. AI may be a powerful force in shaping our social lives, but it is ultimately up to us to determine whether that force strengthens our bonds—or weakens them.

'The great myth of our times is that technology is communication.'

- Libby Larsen

PAUSE & REFLECT

Chapter 4

(AI in Social Life — The Double-Edged Sword?)

Have you ever noticed bias in an algorithm or platform you use?..
....
....
....

What ethical responsibilities should AI creators hold most sacred?
....
....
....
......

Do you believe AI can ever be neutral? Why or why not?
....
....
....
....

'The greatest danger of Artificial Intelligence is that people conclude too early that they understand it.'

-Eliezer Yudkowsky

AI and Decision-Making: Who Do We Trust?

AI: The Decision-Maker

Artificial Intelligence (AI) has become an indispensable part of decision-making in a variety of industries. Its ability to process massive datasets, detect patterns, and deliver insights has revolutionized how choices are made, especially in situations where speed and precision are critical. To better understand AI's role, imagine it as a highly efficient assistant that doesn't rely on intuition or emotions as humans do. Instead, it meticulously analyses data to generate informed recommendations. This capability is particularly beneficial in areas where time-sensitive and data-intensive decisions must be made—tasks that would overwhelm human capacity.

For example, take the world of finance. Investors constantly need to monitor market trends, interpret complex economic data, and predict shifts that could affect their portfolios. Traditionally, this work required hours of painstaking analysis by financial experts. With AI, however, these processes are streamlined. AI systems monitor global markets in real-time, spotting patterns, anomalies, or trends far quicker than any human could. If a disruption in oil prices signals instability in related industries, an AI tool can recommend portfolio adjustments almost instantly. By optimizing financial decisions with speed and accuracy, AI empowers investors to make better choices during volatile economic periods.

Similarly, AI's transformative role in healthcare demonstrates its value as a decision-making tool. Medical professionals, particularly radiologists, are increasingly turning to AI to detect diseases in their early stages. For instance, an AI system trained on thousands of medical images can identify subtle signs of cancer that might go unnoticed by even the most experienced doctor. This level of precision can save lives by ensuring that patients receive timely and accurate diagnoses. Moreover, because AI doesn't suffer from fatigue or cognitive overload, it applies the same focus and consistency to every case, enhancing the overall reliability of medical assessments. Yet, AI's role in healthcare isn't about replacing doctors—it's about enhancing their capabilities, allowing them to combine AI-driven insights with their clinical expertise for better outcomes.

As remarkable as AI's capabilities are, its limitations cannot be ignored. AI's decisions are entirely dependent on the quality of the data it's trained on. Think of data as the foundation of AI's "knowledge."

If this data is flawed, incomplete, or biased, the decisions AI makes will reflect those imperfections. For example, imagine a company using AI in its hiring process. If the system is trained on historical hiring data that disproportionately favours male candidates, it may inadvertently prioritize men over equally qualified women. Similarly, in healthcare, if an AI system is trained on datasets that underrepresent specific demographic groups, its diagnostic accuracy for those populations may be compromised.

Such examples reveal the critical need for scrutiny in the development and deployment of AI. The promise of AI lies in its ability to analyse data on an unprecedented scale, but this power must be wielded responsibly. Developers, organizations, and users need to ensure that the data feeding these systems is diverse, representative, and free of biases. Otherwise, AI risks perpetuating inequalities and making decisions that are neither fair nor inclusive.

This reliance on data also highlights why human oversight is essential. While AI excels at processing information and identifying patterns, it lacks the empathy, ethical judgment, and contextual understanding that humans bring to decision-making. For instance, a radiologist might use AI as a second opinion to confirm their diagnosis but would ultimately rely on their own expertise and empathy to determine the best course of action for the patient. Similarly, financial analysts might validate AI-generated predictions with their own critical thinking, ensuring that the tool's recommendations align with broader market conditions and client needs.

AI is a powerful ally in decision-making, but it is not without its flaws. Its brilliance lies in its ability to handle tasks that are too complex or time-consuming for humans. In finance, it predicts market trends; in healthcare, it saves lives through early detection of diseases. However, its reliance on data—both its strength and its Achilles' heel—underscores the importance of thoughtful implementation and oversight. Decisions that affect people's lives, careers, or well-being require a balance of AI's efficiency and humanity's judgment. By understanding and addressing these complexities, we can ensure that AI serves as a tool that complements human abilities rather than replacing them.

Human Oversight in AI Decision-Making

The relationship between Artificial Intelligence (AI) and human oversight is a complex and evolving one. While AI has demonstrated its ability to make rapid and informed decisions, it is critical that humans remain actively involved in these processes to ensure that decisions align with ethical, social, and contextual considerations. Trusting AI blindly can result in significant risks, from perpetuating biases to undermining fairness and justice. Human oversight acts as the safeguard that bridges the technological capabilities of AI with the values and empathy inherent in human decision-making.

One of the most striking examples of this dynamic is found in the legal system. In certain jurisdictions, AI tools are employed to assess whether a defendant is likely to re-offend—a prediction that can significantly influence bail decisions or sentencing recommendations. On paper, this technology aims to reduce human biases by relying on data-driven analysis. However, studies have shown that these algorithms can themselves exhibit racial prejudice, often due to biased training data. For example, if the historical data used to train the AI disproportionately associates specific communities with higher rates of re-offense, the AI's predictions will reflect and amplify this bias. In such cases, a judge who relies solely on AI recommendations risks perpetuating systemic inequities rather than addressing them. Human oversight becomes essential here—not only to verify the accuracy of these predictions but also to ensure that decisions are made with fairness and justice at their core.

The hiring process offers another striking example of the dangers of unchecked AI decision-making. Many companies use AI-powered screening tools to evaluate job applications, aiming to streamline recruitment by identifying candidates who best match predefined criteria. However, these systems can unintentionally discriminate based on historical biases embedded in the data. One well-documented case involved an AI hiring tool that consistently favoured male candidates over female ones, simply because the training data reflected past hiring practices that prioritized men. This example underscores the importance of human intervention—not only to identify and correct biases but also to champion diversity and inclusion in the workplace. Without human oversight, such systems risk reinforcing inequality under the guise of efficiency.

Human oversight also adds dimensions to decision-making that AI cannot replicate: empathy and contextual understanding. Take healthcare as an example.

While AI systems excel at analysing patient data and recommending treatment options based on patterns and probabilities, they cannot account for emotional or psychological factors unique to each individual. Imagine a scenario in which a patient feels apprehensive about undergoing a particular treatment due to personal fears or past experiences. While the AI may recommend this treatment as statistically optimal, a doctor—considering the patient's emotional state—might propose an alternative approach that prioritizes comfort and trust. This level of understanding is beyond the scope of even the most advanced algorithms.

Additionally, contextual judgment is vital in situations where cultural, social, or ethical nuances come into play. For instance, an AI system might suggest cost-cutting measures for a business without recognizing the potential harm to employee morale or long-term organizational stability. A human leader, on the other hand, would consider these broader implications and weigh them against financial efficiency, ensuring that the decision aligns with the company's values and objectives.

Ultimately, the role of human oversight is not to replace or undermine AI's capabilities but to complement them. While machines offer speed and precision, humans bring ethical judgment, creativity, and emotional intelligence—qualities that cannot be reduced to data points or algorithms. The key to successful decision-making lies in collaboration: combining the strengths of AI with the unique insights of human beings. Whether in law, healthcare, or recruitment, this partnership ensures that decisions are not just accurate but also empathetic, equitable, and reflective of the complexities of the human experience.

By maintaining this delicate balance, we can harness AI's potential while safeguarding against its limitations, creating systems that serve both efficiency and humanity.

The Paradox of AI Trust

The concept of trust in Artificial Intelligence (AI) presents a fascinating paradox. On one hand, AI systems are celebrated for their impartiality, speed, and precision. These qualities allow them to analyse complex data sets and make logical, consistent decisions, often free from the biases or emotional influences that humans might bring to the table. On the other hand, AI lacks the very qualities that make human decision-making uniquely valuable: intuition, emotional intelligence, creativity, and ethical judgment. This tension becomes particularly evident in high-stakes scenarios where the outcomes of decisions can have profound implications.

Consider aviation, where the stakes are literally life and death. Picture this: an airplane is experiencing a mechanical failure, and the AI pilot quickly processes thousands of pieces of data—weather conditions, system diagnostics, and historical records of similar failures. Based on this data, the AI determines the optimal course of action and recommends it. Meanwhile, the human pilot, with years of flight experience and an intimate understanding of how machines behave under stress, senses that the situation demands a different response. Perhaps the pilot's instinct tells them that turbulence might render the AI's course risky, or they feel the need to communicate directly with air traffic control for alternative options. Here lies the paradox: whose judgment do we trust—the AI's, grounded in cold, hard logic, or the pilot's, informed by intuition and expertise? Scenarios like this highlight the need for a collaborative approach in which AI and human insights enhance rather than compete with each other.

The same paradox extends into creative industries, where AI is increasingly being used to generate art, music, and designs. AI systems like DALL-E or DeepArt can create stunning visuals or compose intricate melodies that rival human creations in technical complexity. However, the emotional depth and imaginative spark that make human artistry truly exceptional are often absent in AI's output. For example, a designer might use AI to generate initial concepts for a project, benefiting from the tool's ability to process vast creative references and suggest innovative styles. Yet, when it comes to selecting which concept resonates most deeply, the designer will inevitably rely on their own intuition, emotions, and personal experience.

A logo or a piece of music may meet technical perfection through AI, but only a human can imbue it with the unique depth that comes from lived experience and creativity.

The tension between trusting humans and trusting machines is further complicated by the nature of AI itself. Machines operate with logical consistency, but they lack contextual understanding and adaptability. While humans may rely on "gut instincts" in certain situations, AI cannot. For instance, in medicine, an AI system might recommend an aggressive treatment plan for a patient based purely on statistical outcomes. A doctor, however, might consider factors that the AI overlooks—such as the patient's anxiety about invasive procedures or their need for emotional support during recovery. Trusting AI exclusively in this context might lead to decisions that are medically sound but lack compassion and nuance.

The challenge, therefore, is not to pit AI against human decision-making but to foster trust in the collaboration between the two. Machines bring logic, efficiency, and analytical rigor to the table, while humans contribute ethics, empathy, and creativity. The key to navigating this paradox lies in developing clear frameworks that define the boundaries of AI's decision-making authority and specify when and how human input should intervene.

For example, industries that rely heavily on AI need to establish guidelines for validation and oversight. In aviation, protocols could ensure that AI's recommendations are always cross-checked by pilots. In creative fields, human creators might use AI as a brainstorming partner but remain the final arbiters of artistic choices. In healthcare, doctors might treat AI suggestions as advisory, relying on their own expertise to make the ultimate decision.

By fostering collaboration rather than competition between humans and machines, we can leverage the strengths of both while mitigating their weaknesses. Trust in AI does not require relinquishing control to algorithms; rather, it requires understanding their capabilities and limitations and integrating them thoughtfully into decision-making processes. The goal is not to choose between humans and machines but to build a future in which their partnership produces decisions that are not only efficient but also ethical, empathetic, and deeply human.

Ethics and Accountability

The integration of Artificial Intelligence (AI) into decision-making processes has undoubtedly revolutionized industries, offering efficiency and precision that were previously unattainable. However, this technological leap comes with profound ethical questions, particularly when errors occur. Who should be held accountable for the consequences of AI decisions? Is it the machine itself, the developers who designed it, the organization deploying it, or the user relying on it? This question becomes even more critical in scenarios where the stakes are high and the impact of a mistake can lead to harm, loss, or controversy.

Take the example of self-driving cars. These vehicles rely on AI to make split-second decisions about safety. Whether it's recognizing a pedestrian crossing the street, determining the best evasive manoeuvre in an accident scenario, or stopping for an unexpected obstacle, AI must process real-time data and act quickly. But what happens when a self-driving car fails to respond appropriately, leading to a crash? Determining accountability in such cases becomes incredibly complex. Was the failure caused by a flaw in the programming, an error in the data the car was trained on, or negligence from the user who was supposed to monitor the car's behaviour? Without clear frameworks for accountability, the repercussions of AI errors risk falling into a grey area, leaving victims without closure and developers, companies, and users unsure of their roles and responsibilities.

The finance industry provides another compelling example of the ethical dilemmas surrounding AI decision-making. AI tools used for financial analysis are designed to interpret market data, identify trends, and generate recommendations for investments or risk mitigation. But these systems are not infallible—errors in data interpretation can lead to significant losses. Imagine an AI misinterpreting economic indicators and recommending a course of action that results in a company losing millions of dollars. Who is responsible for the mistake? Should the blame fall on the developers who created the tool, the analysts who relied on it without verifying its output, or the executives who decided to deploy it at scale?

These questions of accountability become even more pressing when considering the power AI holds in sensitive areas like justice, healthcare, and hiring. In the legal system, for instance, AI tools are used to predict sentencing outcomes or assess the likelihood of re-offense.

If such tools make biased judgments due to flaws in their training data, the repercussions could undermine justice itself. Similarly, in hiring processes, AI systems may unknowingly perpetuate discrimination based on gender or race, despite their intended purpose of streamlining recruitment. Ethical accountability in these situations demands answers to difficult questions: Should developers be held liable for not addressing bias in their systems? Or should companies bear responsibility for deploying flawed AI tools without proper oversight?

Transparency is a cornerstone of ethical AI practices. Without a clear understanding of how AI systems make decisions, users and stakeholders are left in the dark, unable to evaluate the fairness or reliability of the outcomes. For instance, self-driving car manufacturers must clearly explain how their vehicles' AI systems are programmed to prioritize decisions in dangerous situations—such as whether to swerve to avoid a pedestrian at the cost of endangering passengers. In finance, companies deploying AI tools must disclose how these systems interpret market data, ensuring clients understand their methodologies and limitations.

The demand for explainability in AI systems is critical to building trust. Users need to understand the mechanisms behind AI decision-making, from the data it relies on to the algorithms that process it. Safeguards must be implemented to mitigate risks, such as regular audits, bias testing, and the inclusion of fallback mechanisms for manual intervention when errors occur. Ethical AI practices also require collaboration between developers, regulators, and end-users, creating standards that balance innovation with accountability.

The absence of such measures undermines trust in AI systems. If the public perceives AI as opaque, uncontrollable, or biased, its adoption will face resistance, regardless of its potential benefits. Addressing this requires proactive steps toward transparency and accountability, ensuring that developers and companies are held responsible for their creations, and users are informed enough to use AI systems responsibly.

Ultimately, the ethical challenges surrounding AI decision-making boil down to ensuring fairness, accountability, and trust. Clear frameworks for responsibility are necessary, not just to address errors when they occur but to prevent them in the first place. By prioritizing transparency and collaboration, we can ensure that AI systems serve as tools of progress rather than sources of harm. These measures are vital to building a future where AI decisions are reliable, equitable, and ultimately supportive of humanity's best interests.

The Role of Bias

Bias is one of the most pressing and complex challenges facing Artificial Intelligence (AI) today, particularly in decision-making processes. At its core, AI operates by analysing patterns in data and using these patterns to predict outcomes or make recommendations. While this may seem efficient and impartial, the problem arises when the data used to train these algorithms reflects inequalities or biases embedded in society. Instead of providing fair and unbiased decisions, AI systems can inadvertently perpetuate discrimination, injustice, or systemic inequities that their creators might not have intended.

To understand bias in AI, it's important to consider how these systems learn and make decisions. AI algorithms are trained on historical data—a collection of information from past events or trends. While this data serves as the foundation for AI's "knowledge," it carries with it the prejudices, disparities, and unequal treatment that exist in the real world. For example, if historical hiring data shows a preference for male candidates over female candidates, an AI trained on this data may prioritize men in its recruitment recommendations, unintentionally reinforcing this bias. This creates a loop where past inequalities are repeated instead of corrected.

Bias in AI isn't limited to hiring. In real estate, AI tools are used to predict property values, assess risks, and recommend investment strategies. But what happens when the training data includes patterns of segregation or redlining—the discriminatory practice of denying services to certain communities based on racial demographics? The AI, unknowingly learning from this data, may suggest investment strategies that continue to exclude those communities, perpetuating the inequities its creators hoped to address. Similarly, in healthcare, AI systems designed to recommend treatments rely on datasets containing medical records and demographic statistics. If these datasets underrepresent specific groups—for instance, women or minority populations—the AI might prioritize treatments for demographic groups with higher representation in the data. This can lead to poorer healthcare recommendations for underrepresented groups, highlighting disparities between AI's capabilities and its fairness.

The consequences of bias in AI decision-making are far-reaching and often amplify existing inequalities. Injustice can become more pronounced, as biased algorithms reinforce systemic inequities instead of eliminating them. In the legal system, AI tools are used to predict re-offense rates, which can influence sentencing or bail decisions. If the training data associates certain racial groups with higher crime rates, the AI may recommend harsher penalties for individuals from those communities—an outcome that exacerbates societal divisions and undermines trust in the justice system. In finance, biased AI algorithms could deny loan approvals to individuals from historically underserved communities, basing these decisions on patterns from past data rather than individuals' actual creditworthiness. This perpetuates cycles of economic inequality, making it harder for marginalized groups to access resources and opportunities. Such examples underscore why addressing bias is critical—not just for ethical reasons but to ensure that AI systems deliver equitable, reliable, and effective outcomes.

Fixing bias in AI systems is a challenging task, but it is not impossible. One essential measure is ensuring diverse representation in AI development teams. Developers from varied backgrounds and perspectives are more likely to identify and address biases that might otherwise go unnoticed. For example, a team that includes members from historically marginalized communities may be better equipped to recognize disparities in training data and advocate for corrections. Beyond the development stage, rigorous testing of algorithms before deployment is crucial. Developers must actively seek out biases during the testing phase, evaluating how the system performs across different demographics, conditions, or scenarios. If disparities are identified—such as an AI favouring one group over another—corrective measures must be taken to ensure fairness.

Bias auditing tools can also play a significant role in identifying and mitigating bias in AI systems. These tools analyse algorithms to identify and quantify biases, providing developers with actionable insights for refinement. For example, a hiring AI that disproportionately excludes women could be adjusted after an audit reveals its bias, ensuring that the system evaluates candidates equitably. Transparency in AI systems is equally important. By designing AI tools with explainability in mind, developers can help users understand how decisions are made. If an AI tool denies a loan application, it should explain its reasoning—whether based on income level, credit history, or other factors. This transparency allows users to identify potential biases and hold developers accountable for addressing them.

Ultimately, eliminating bias in AI is about promoting equity rather than reinforcing inequality. Developers must prioritize fairness at every stage, from training data collection to algorithm design and deployment. Policymakers, ethicists, and organizations must collaborate to create standards that ensure AI systems are not just effective but also equitable. By actively addressing bias, we can unlock AI's potential to serve as a force for good—helping reduce disparities, expand opportunities, and improve lives. However, achieving this goal requires vigilance, accountability, and a commitment to building systems that reflect the diversity and fairness we seek in society. Bias is not an insurmountable obstacle, but it demands thoughtful, proactive action to overcome. By taking these steps, we can ensure that AI systems elevate all communities rather than deepening divides.

The Way Forward

As we reach the conclusion of this chapter on AI and decision-making, it becomes clear that Artificial Intelligence is both a transformative force and a complex challenge. Its ability to process vast amounts of data and deliver insightful recommendations has changed how we make decisions, enabling progress across industries like healthcare, finance, and justice. Yet, AI's potential is only as impactful as the framework within which it operates. To create a future where AI complements human decision-making rather than undermining it, thoughtful integration is key.

Collaboration lies at the heart of this integration. AI's strengths—efficiency, precision, and impartiality—must be harmonized with human capabilities like creativity, ethical judgment, and intuition. Machines can provide logic-driven recommendations, but humans possess the ability to evaluate those decisions in nuanced and meaningful ways. For example, while an AI algorithm might propose a course of action based on statistical likelihood, a human decision-maker can consider broader contextual factors, emotional dynamics, and ethical implications that algorithms cannot grasp.

Education is an essential part of achieving this balance. Users need to understand both the capabilities and limitations of AI to make informed decisions about when to rely on machines and when to trust human instincts.

For instance, professionals using AI in healthcare might undergo training to interpret AI-generated diagnoses alongside their clinical expertise, ensuring that technology serves as a complement rather than a replacement. Similarly, investors using AI-driven financial tools should be aware of the inherent risks of relying solely on algorithms, recognizing when to validate AI recommendations with human judgment.

Policymakers and developers also play a crucial role in shaping the future of AI decision-making. Collaboration between these stakeholders is necessary to create guidelines that prioritize fairness, accountability, and transparency. For instance, regulatory frameworks could mandate bias testing during the development of AI systems, ensuring that algorithms are thoroughly examined for disparities before deployment. Developers could be required to document how their systems reach decisions, allowing users to understand and challenge outcomes when necessary. These steps not only enhance the reliability of AI systems but also foster public trust in their implementation.

Transparency is particularly vital in building this trust. AI systems should be designed with explainability in mind, enabling users to understand the logic behind their decisions. Imagine an AI tool recommending a loan rejection—it should clearly outline whether the decision was based on credit history, income, or other factors. Such transparency empowers users to question decisions, identify biases, and hold systems accountable. Without this openness, trust in AI remains tenuous, limiting its adoption and effectiveness.

Ultimately, AI should not replace human decision-making—it should amplify it. By fostering trust in both humans and machines, we can create a future where decisions are not only efficient but also ethical and empathetic. For example, consider a legal system where judges use AI tools to analyse case histories and predict outcomes but rely on their own judgment to account for individual circumstances and ethical considerations. Or imagine a creative industry where AI generates innovative concepts, but human artists infuse them with emotion, personality, and depth.

These partnerships illustrate how AI and human intelligence can work hand in hand, producing outcomes that honour both technological progress and human values.

The way forward requires vigilance, accountability, and a commitment to equity. Bias must be actively confronted, and ethical frameworks must be continuously refined as AI evolves.

Developers, users, policymakers, and ethicists must work together to ensure that AI systems reflect the diversity, fairness, and humanity we aspire to. Trust in AI is not about relinquishing control—it is about fostering collaboration, building systems that serve as allies rather than adversaries in decision-making.

Artificial Intelligence has the potential to redefine how we approach decisions, transforming industries and improving lives. But this transformation is only possible if we integrate AI thoughtfully and responsibly, preserving the qualities that make human decision-making irreplaceable. The future of AI is not just a technological one; it is a deeply human journey—one that invites us to balance logic with empathy, efficiency with creativity, and innovation with equity. By embracing this balance, we can ensure that AI truly serves humanity, creating a world where decisions are not just smart but deeply meaningful.

'AI doesn't have to be evil to destroy humanity—if AI has a goal and humanity just happens to stand in the way, it will destroy humanity as a matter of course without even thinking about it.'

-Elon Musk

PAUSE & REFLECT

Chapter 5

(AI in Social Life — The Double-Edged Sword?)

Have you ever noticed bias in an algorithm or platform you use?...
....
....
....

What ethical responsibilities should AI creators hold most sacred?
....
....
....
......

Do you believe AI can ever be neutral? Why or why not?
....
....
....
....

Ethics in AI:

The Dark Side of Intelligence

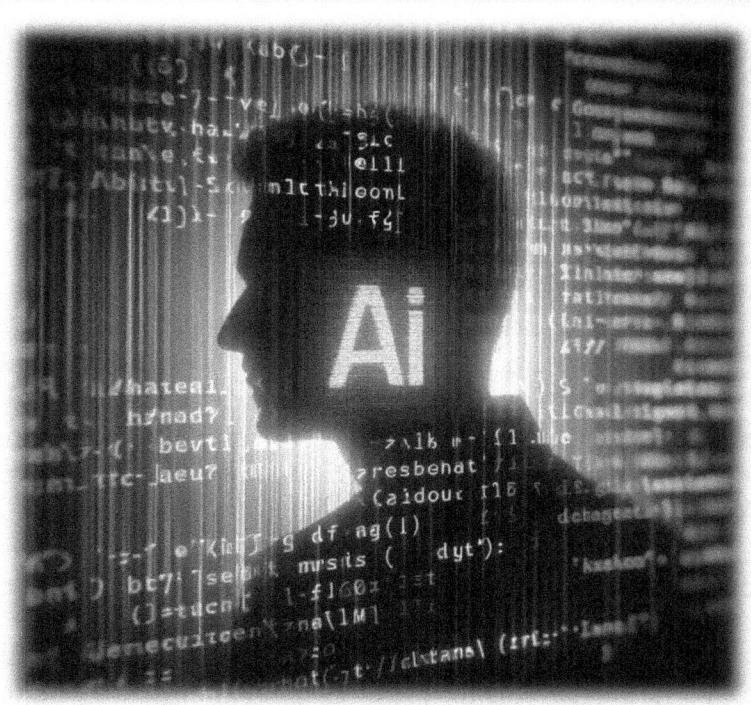

Artificial Intelligence (AI) stands as one of humanity's most extraordinary achievements—a technological marvel capable of transforming industries, streamlining complex processes, and improving lives on an unprecedented scale. With its ability to process immense amounts of information in seconds, automate decisions, and uncover patterns that elude human comprehension, AI has driven revolutionary progress across domains like healthcare, finance, education, and beyond. Yet, this extraordinary power is not without risk. AI, while seemingly impartial and objective, is not immune to the flaws embedded in society. Like a mirror, it reflects the intentions, decisions, and biases of its creators, magnifying both our best qualities and our deepest shortcomings.

At its heart, AI learns from us—from the data we feed it, the systems we design, and the values we program into it. But herein lies the danger: when societal biases and inequalities seep into AI's algorithms, the technology not only reproduces these flaws but amplifies them. A biased hiring algorithm, for example, can deny opportunities to entire groups of people, reinforcing prejudices that should be dismantled. A flawed surveillance system can unfairly target communities already vulnerable to discrimination. Rather than eliminating these inequities, AI risks entrenching them even further, raising profound ethical questions about fairness, accountability, and the protection of human rights.

Ethics in AI is no longer a peripheral concern relegated to academic debate or niche tech discussions. It is a defining issue—central to how AI shapes our social, political, and individual lives. Algorithms that determine who gets hired, promoted, or fired wield immense influence. Systems that guide healthcare decisions or judicial outcomes impact millions of people. And yet, the ethical dilemmas surrounding AI remain unresolved. Who decides what is fair? Should AI systems be entrusted with decisions that can alter lives? And what happens when this powerful technology is misused—to manipulate, monitor, or oppress?

These questions are not hypothetical—they are unfolding in real-time. From biased hiring systems to AI-powered misinformation campaigns, the misuse of AI is causing ripple effects across the globe. Predictive algorithms in law enforcement are reinforcing systemic injustices, AI surveillance systems are infringing on privacy and human rights, and deepfake technologies are undermining trust in media and individuals alike. Yet, alongside these challenges, AI has also demonstrated its potential to drive progress.

It is diagnosing diseases, combating climate change, and enhancing global communication. Its power to uplift humanity is as real as its ability to harm, making ethics in AI not just a technological issue but a profound reckoning with the choices we make and the values we uphold.

As we dive into this chapter, we will explore the complex ethical landscape of AI, dissecting real-world cases where it succeeded and where it failed. We will examine biased systems, manipulative algorithms, and surveillance technologies to understand where AI went wrong—and we will highlight moments of progress where AI was harnessed responsibly for the greater good. Central to these discussions will be the principles of inclusive data, transparent design, and moral accountability—all necessary for guiding AI toward ethical outcomes. But more importantly, this chapter will reflect on the human role in shaping AI's path. Whether AI becomes a force for empowerment or exploitation is not determined by the technology itself—it is determined by us.

AI's journey is intertwined with our own. It is not just a tool—it is a mirror of our humanity, capable of magnifying our ambitions and exposing our flaws. In exploring ethics in AI, we are compelled to confront the values we embed in our systems, the vigilance we exercise in their implementation, and the humanity we bring to every decision. This chapter invites you to consider the stakes of this powerful technology, asking the question: Will AI reflect our highest ideals or our worst instincts? The answer lies in our choices. The future of intelligence begins here.

A Mirror of Our Makers: The Inherited Flaws of AI

AI systems are designed to learn from data, but they do so without understanding the social and cultural contexts in which that data is embedded. This is both their greatest strength and their greatest vulnerability. Consider a hiring algorithm designed to recommend candidates for a job. On the surface, it might seem objective, basing its decisions purely on qualifications, experience, and performance metrics. However, if the data used to train this algorithm reflects historical biases—favouring certain genders, ethnicities, or socio-economic backgrounds—those biases will become part of the AI's logic. The result? An ostensibly fair system that reinforces existing inequalities.

One striking example occurred in 2018, when a prominent tech company abandoned its AI hiring tool after discovering it discriminated against women. The AI was trained on ten years of internal hiring data—a dataset that favoured male candidates due to historic trends in male-dominated industries. Instead of neutralizing bias, the system penalized resumes that included the word "women's" (as in "women's chess club captain") or attended all-female colleges. The lesson is clear: AI is not inherently neutral—it is a reflection of the data it is given and the choices of its creators.

This problem extends beyond hiring. In real estate, AI tools used to predict property values or recommend development areas have been found to reproduce patterns of segregation and discrimination. If the training data includes decades of redlining—a practice where minority communities were systematically denied loans or investment—the AI learns to perpetuate those inequities. In these cases, technology becomes a tool not of progress, but of entrenching historical injustices.

The bias in AI is not just a technological issue—it's a human one. Developers and organizations must take responsibility for interrogating the data they use and the assumptions built into their systems. Without this ethical vigilance, AI risks becoming a magnifying glass for humanity's worst traits.

Who Gets to Decide What's Fair?

Fairness is a concept as old as human society, yet it remains deeply subjective and context-dependent. What seems fair to one group might feel profoundly unjust to another. It is shaped by cultural norms, social structures, historical experiences, and individual values. When Artificial Intelligence (AI) enters the equation, the challenge of defining fairness becomes even more complex, as algorithms attempt to codify and apply abstract concepts into actionable decisions. This raises a critical question: who gets to decide what fairness looks like in AI systems? Is fairness about equal treatment—offering everyone the same opportunities—or is it about equity, which involves recognizing and addressing systemic disadvantages to level the playing field?

The difficulty in defining fairness lies not in a lack of intention but in its profound subjectivity.

For example, what is "fair" in the context of hiring? Should AI systems focus on a merit-based model, prioritizing qualifications and experience above all else? Or should they account for socio-economic barriers that may have limited opportunities for certain individuals, such as unequal access to education? Fairness in hiring could mean advocating for candidates from underrepresented groups to achieve equity in representation, but for others, this might conflict with the principle of equal treatment. When humans struggle to define fairness, the task of programming AI to uphold fairness becomes exponentially more challenging.

This complexity is highlighted in the example of predictive policing algorithms—a real-world case where fairness has been hotly debated. These systems analyse historical crime data to identify "high-risk" neighbourhoods or individuals for law enforcement intervention. The intention is to make policing more efficient, reducing crime by allocating resources where they're most needed. But the reality is far more complicated. If the input data reflects biased practices—such as the over-policing of certain communities or demographic groups—the algorithm will replicate and reinforce these biases. Imagine a neighbourhood that historically had higher crime reports not due to actual crime rates but because of disproportionate police scrutiny. Predictive policing algorithms trained on this data might classify the neighbourhood as "high risk," perpetuating racial profiling and systemic injustice. This raises the question: is it fair to use AI to predict crimes when the very concept of "risk" is tied to flawed and biased data?

Healthcare poses another challenging dilemma when it comes to fairness. AI systems in medicine often prioritize treatments based on statistical outcomes, recommending interventions for populations with higher recovery rates or better cost-benefit ratios. At first glance, this seems logical—maximizing positive outcomes by allocating resources to those most likely to benefit. However, this approach can unintentionally sideline marginalized groups who need care the most. For instance, individuals from economically disadvantaged backgrounds may have lower recovery rates, not because of inherent health risks but due to limited access to healthcare, nutritious food, or safe living conditions. If AI systems fail to account for these broader social contexts, they risk perpetuating inequality in healthcare access and outcomes.

In this case, fairness isn't just about efficiency—it's about ensuring that care reaches those who are most vulnerable.

These dilemmas reveal an uncomfortable truth: without a clear and shared understanding of fairness, AI systems are at risk of becoming tools that amplify the biases of those in power. An algorithm programmed by individuals with privileged perspectives may unintentionally exclude the voices and needs of marginalized communities. For example, in hiring, healthcare, or law enforcement, AI might prioritize outcomes that align with the worldview of developers or organizations, failing to address or even exacerbating systemic inequalities.

To ensure ethical outcomes, fairness must be defined inclusively, taking into account diverse perspectives and lived experiences. The design process for AI systems should involve a broad range of voices—from developers and ethicists to members of historically marginalized communities. For instance, an AI development team designing a hiring algorithm might consult with diversity advocates to better understand the barriers faced by minority groups. Policymakers could create frameworks requiring predictive policing algorithms to undergo bias audits before deployment, ensuring that they do not disproportionately target certain demographics.

Transparency and accountability are also critical. AI systems should make their decision-making processes explainable, so users can challenge outcomes and identify potential biases. For example, if a healthcare AI prioritizes one population over another, it should provide a detailed rationale—whether based on recovery rates, cost-effectiveness, or other factors. This transparency empowers stakeholders to question and refine the criteria that guide decisions, fostering trust and ensuring that AI systems are held accountable for their impact.

In the end, the question of fairness in AI is not just a technological challenge—it is a profoundly human one. Fairness is not static; it evolves as societies change and grow. AI systems must be designed to adapt to these shifts, ensuring that they remain tools for progress rather than instruments of exclusion. The responsibility to define fairness lies with everyone involved—developers, policymakers, organizations, and citizens. By engaging diverse voices, prioritizing transparency, and creating ethical frameworks, we can guide AI toward outcomes that reflect equity, inclusion, and humanity.

Fairness in AI is not a one-size-fits-all solution, nor can it be achieved through simple programming alone. It is a continuous process of reflection and refinement, shaped by the values we bring to the systems we create.

As AI becomes increasingly integrated into decision-making processes, the challenge of defining fairness will only grow. But with vigilance, collaboration, and ethical intention, we can ensure that AI reflects the best of humanity rather than its worst instincts. In answering the question "Who gets to decide what's fair?" we are also deciding who gets to shape the future—and the stakes couldn't be higher.

When AI Misleads, Manipulates, and Monitors

The ethical dilemmas surrounding Artificial Intelligence (AI) often feel like the tension in a tightly wound thread, ready to unravel into either progress or peril. Perhaps no issue highlights this tension more dramatically than the potential for AI to mislead, manipulate, and monitor. What was once a tool designed to enhance human lives can, in the wrong hands or under insufficient oversight, become a force that infringes on personal freedoms, erodes democratic values, and threatens the very principles of human dignity.

Consider the use of AI-powered surveillance systems, a troubling example of technology's capacity to serve both the powerful and the oppressive. In some authoritarian regimes, AI surveillance has become a cornerstone of controlling populations. Facial recognition technology, for instance, is used to track citizens, monitor public spaces, and identify dissenters. These systems do not discriminate when programmed by humans with intent to suppress: they can target minorities, activists, or simply those who step out of line with the state's ideology. While supporters may argue that such tools enhance public safety, the price is often paid in the form of lost privacy and diminished freedoms. What does it mean to live under constant observation, where every step and glance is recorded, judged, and potentially weaponized against you? The implications for human rights are staggering.

Even in democratic societies, AI can influence behaviour in subtler yet no less concerning ways. Take social media algorithms, for example.

These systems analyse user data to provide personalized content, a process that seems harmless—helpful, even—on the surface. But these same algorithms have a darker side: they often amplify divisive content to maximize engagement. During election campaigns, AI tools have been used to micro-target voters with misleading ads or outright fake news.

The result is not just a misinformed electorate but a fractured society. Social media becomes an echo chamber where opinions are reinforced, opposition is demonized, and dialogue disintegrates. Here, the question shifts from whether AI can influence behaviour to how we ensure that influence is responsible and ethical.

Perhaps one of the most unsettling forms of AI misuse is the exploitation of deepfake technology. Deepfakes—hyper-realistic fake videos generated using AI—are no longer just the stuff of science fiction. They have been used to create convincing but entirely false narratives, manipulate political outcomes, and harass individuals. Imagine the psychological trauma of having your face superimposed onto explicit or humiliating content, as was the case with a journalist who fell victim to this malicious use of AI. Such instances don't just harm individuals— they corrode the foundation of trust in society. When we can no longer distinguish between real and fabricated evidence, what happens to our ability to hold the powerful accountable or believe in truth itself?

The common thread in all these examples—whether surveillance, manipulation, or deepfakes—is not the technology itself but the intent behind its use. AI, after all, is a tool; its morality is shaped by the choices of its makers and users. Yet this does not absolve us of responsibility. The dangers of misuse make it clear that robust safeguards are essential. Governments, international organizations, and tech companies must work together to create regulatory frameworks that establish ethical boundaries for AI use. Transparency is key: developers must ensure that systems are explainable and auditable, and users must be empowered to challenge and question their decisions.

International cooperation is equally critical, especially as AI knows no borders. The misuse of AI in one part of the world has ripple effects everywhere, whether it's the erosion of democratic norms or the global spread of manipulated information. Laws and guidelines alone, however, are not enough. A cultural shift is needed—one that prioritizes privacy, human rights, and accountability above profit or convenience. Safeguarding human dignity in the age of AI is not an option; it is a moral imperative.

AI has the power to influence our actions, perceptions, and even beliefs. But it is our vigilance, ethics, and humanity that will determine whether AI enhances our freedom or endangers it. By addressing these challenges head-on, we can transform AI from a source of anxiety into a force for good—a tool that serves society while respecting the sanctity of individual rights.

Let us not wait for more harm to unfold before taking action. Let us imagine and demand a future where intelligence, artificial or otherwise, always aligns with the principles of human dignity and justice.

The Path to Ethical AI: Principles and Practices

The journey to creating ethical AI starts not with machines, but with human values—transparency, accountability, and inclusivity. These foundational principles serve as the compass for AI development, ensuring that technology not only functions efficiently but aligns with the moral fabric of society. At its best, AI has the potential to transform lives, build bridges, and tackle complex global challenges. But at its worst, it risks becoming a tool that deepens divides and exacerbates inequality. The path it takes depends on the guiding framework we establish today.

Transparency is perhaps the most vital cornerstone of ethical AI. For AI systems to gain trust, users must understand how they work and how decisions are made. Take, for example, an AI-powered tool that denies a user a mortgage application. Without transparency, the applicant is left in the dark, unable to determine whether the decision was based on their credit history, income level, or some unknown factor hidden within the algorithm. This opacity breeds distrust and leaves users powerless to contest or understand the outcome. Instead, ethical AI systems must be designed to be fully explainable, providing clear insights into the decision-making process. Transparency empowers individuals to hold systems accountable and ensures that AI operates in service of human fairness rather than dominance.

Moreover, transparency fosters collaboration between developers and end-users, bridging the gap between complex technologies and the people they impact. Consider AI in healthcare. When patients or doctors can understand the rationale behind AI-driven diagnoses or treatment recommendations, they

are better equipped to make informed decisions. Patients can trust the system and contribute their own insights, while doctors can validate and refine AI suggestions. This mutual understanding enhances not only the accuracy of decisions but the dignity of those affected by them.

Accountability goes hand-in-hand with transparency, yet it carries an additional layer of responsibility. Developers, organizations, and stakeholders must be willing to own the consequences of the decisions their AI systems make. Accountability means accepting not just praise for successes, but responsibility for failures, biases, and errors. For instance, developers of facial recognition software must conduct rigorous bias audits to ensure their systems perform equitably across diverse demographic groups. If discrepancies are discovered—such as the software misidentifying individuals based on race or gender—accountability demands that the findings be addressed and rectified before deployment. Regular reports on performance, audits, and fixes should be publicly shared, building trust through openness and commitment to improvement.

Accountability also involves recognizing the long-term societal impact of AI. An organization deploying AI for predictive policing, for example, must grapple with the ethical implications of reinforcing systemic biases or infringing on civil liberties. Accountability in this context requires organizations to continuously evaluate the consequences of their systems and to make changes when AI inadvertently harms the communities it aims to serve. In this way, accountability keeps technology in check, aligning it with humanity's collective good.

Inclusivity, the third pillar of ethical AI, is no less essential. AI systems do not operate in a vacuum; they interact with the diverse and varied realities of human life. Ensuring these systems reflect those realities requires representation. Diverse development teams—including individuals from historically marginalized communities—bring critical perspectives to the design process, identifying biases and blind spots that homogeneous teams might overlook. Inclusivity leads to richer, more equitable solutions, where AI serves all communities rather than privileging a select few.

For example, an AI model designed to assess loan applications might inadvertently favour applicants from urban areas over rural ones due to data disparities. A team member familiar with rural financial challenges could flag this bias during development, prompting the creation of a more balanced algorithm. Similarly, developers working on an AI healthcare system might include patient advocates to ensure the technology prioritizes accessibility for underserved populations. Inclusivity does not just enrich the technical output—it strengthens the moral integrity of AI itself.

Inclusivity must also extend to the stakeholders affected by AI systems. Engaging with users, advocates, and communities ensures that the technology reflects diverse needs and expectations. For instance, involving educators in the design of AI tools for personalized learning can help ensure these systems adapt to different cultural contexts and learning styles. Such collaboration creates solutions that are as human-centred as they are technologically advanced. To walk the path of ethical AI is to embrace the responsibility of shaping intelligence into a force for good. This is not simply about programming; it is about values, collaboration, and vigilance. Transparency builds trust, accountability ensures responsibility, and inclusivity fosters equity. Together, these principles form the foundation of a future where AI serves humanity— uplifting rather than exploiting, empowering rather than oppressing.

But this path is neither simple nor fixed. The ethical landscape of AI is constantly evolving as technology advances, and with it, so must our principles and practices. As we build these systems, we must ask ourselves not only how they function but what they reflect. Are we creating tools that reinforce our highest ideals, or are we embedding the flaws we hoped to overcome? Ethical AI is more than a framework; it is a commitment—a commitment to making intelligence truly intelligent in its service to humanity. Let us make it one worth keeping.

AI's Fork in the Road: Uplifting or Oppressing?

At the heart of Artificial Intelligence (AI) lies a pivotal crossroads—one path leading to empowerment and progress, the other to exploitation and oppression. The choice between these paths is not written into the algorithms themselves but etched into the intentions and decisions of those who create and deploy them. AI is an immensely powerful technology, capable of transforming lives, solving pressing global challenges, and redefining how we interact with the world. Yet, this same power can be wielded to manipulate, exploit, and suppress, making the ethical stewardship of AI not just a responsibility but a moral imperative.

The uplifting potential of AI is boundless. Imagine a world where AI revolutionizes healthcare, enabling early disease detection and personalized treatments that save millions of lives.

Picture education systems tailored to each student's needs, breaking barriers of accessibility and empowering learners in underserved communities. Consider AI-driven solutions for climate change—optimizing renewable energy sources, tracking deforestation, or predicting natural disasters to mitigate their impact. In these scenarios, AI becomes a beacon of progress, a tool that amplifies humanity's capacity to address its greatest challenges.

But on the flip side lies the darker potential of AI—a potential marked by misuse and ethical neglect. When AI is deployed to manipulate elections, erode the privacy of populations, or reinforce systemic biases, its promise turns into peril. Algorithms designed to sway public opinion with misinformation, facial recognition systems used to target and survey marginalized communities, and hiring tools that perpetuate inequality—all of these examples reveal how AI, when left unchecked, can magnify injustice rather than mitigate it. In such cases, AI ceases to be a tool for progress and becomes an instrument of exploitation, reflecting not humanity's ideals but its flaws.

This fork in the road is not a technological inevitability—it is a human one. The trajectory AI takes depends entirely on the choices made by developers, policymakers, organizations, and citizens. As developers, the responsibility is to build systems that prioritize fairness, accountability, and transparency. As policymakers, the duty is to craft frameworks that safeguard human rights and ensure ethical compliance. As organizations, the charge is to deploy AI responsibly, aligning its outcomes with societal well-being rather than corporate gain. And as citizens, the role is to stay vigilant, questioning and challenging the systems that shape our lives.

Ethical principles must guide every decision, every algorithm, and every application. Fairness must be woven into the fabric of AI systems, ensuring they empower rather than exclude. Inclusivity must be championed, so that diverse perspectives shape how AI interacts with the world. Accountability must be upheld, with mechanisms to address biases, rectify errors, and hold creators responsible. But above all, humanity must remain central—not as an abstract value but as the guiding force behind every choice.

The decisions we make today will shape AI's legacy for generations to come. Will AI be remembered as the technology that lifted humanity to new heights, or as the one that cast shadows of exploitation over our shared future? This chapter is not just a critique of AI's ethical challenges—it is a call to action, a reminder that intelligence, whether artificial or human, must always serve the greater good.

Let the choices we make be guided by fairness, inclusion, and an unwavering belief in human dignity. Let them reflect our collective aspiration to build a future where technology is a force of empowerment and progress, rooted in ethics and humanity. In this fork in the road, the responsibility lies not with AI itself but with us. The path AI takes is ours to choose, and the stakes have never been higher. Let us choose wisely.

'Technology is a useful servant but a dangerous master.'

-Christian Lous Lange

PAUSE & REFLECT

Chapter 6

(Ethics in AI:

The Dark Side of Intelligence?)

Would you trust an AI to make a decision that significantly impacts your life? Why or why not?
....
....
....

What values do you think are essential to embed in AI systems- and who should decide those values?
....
....
....
......

Can you recall a time when a decision made by a machine seemed unfair? How would you redesign it to be more just?
....
....
....
....

'The question of whether a computer can think is no more interesting than the question of whether a submarine can swim.'

- Edsger W. Dijkstra

Chapter 7

Redefining Humanity: What Does It Me to Be Intelligent?

Intelligence is the quality we have long associated with creativity, adaptability, problem-solving, and imagination—the very traits that have set humanity apart from the natural world. For centuries, intelligence was seen as the crowning feature of human existence, the ability that propelled societies forward and unlocked the mysteries of the universe. But now, Artificial Intelligence (AI) is rewriting this narrative, forcing us to rethink not only what intelligence means but also what it means to be human. Is intelligence purely a matter of computation and logic? Or does it extend into realms of emotion, intuition, and the ineffable spark of creativity?

As machines begin to replicate human capabilities—and, in some areas, surpass them—the concept of intelligence becomes increasingly fluid and complex. AI's ability to compose music, paint masterpieces, or generate poetry challenges long-held notions of creativity as a distinctly human attribute. The rise of generative AI is not just a technological marvel; it's a philosophical earthquake, shaking the foundations of identity and humanity itself.

This chapter explores how AI's advancements provoke existential questions about the nature of intelligence. Can machines ever truly understand? Will consciousness emerge from code, or is emotion the final frontier that separates humans from machines? Perhaps even more provocatively, we'll explore transhumanism—the belief that humans and AI might merge to enhance both, blurring the lines between human and machine forever. These philosophical considerations are not abstract; they shape how we approach education, cultural values, and our vision for the future. To redefine intelligence, we must confront the essence of our own.

The Age of Machine Intelligence: Breaking Barriers

Artificial Intelligence has already shattered barriers we once considered unbreachable. Its ability to solve complex problems, analyse massive datasets, and automate decisions has become commonplace in industries worldwide. But beyond these feats of logic lies something more startling: AI's growing proficiency in creativity.

Generative AI systems like OpenAI's GPT models can write poetry that rivals human complexity, emulating the tones, rhythms, and linguistic finesse of history's great authors.

In one stunning example, AI produced Shakespearean sonnets that were so stylistically consistent with the Bard's work that even literary experts were impressed. These systems can create stories, mimic voices, and explore themes with astonishing accuracy. Yet, something remains elusive—something intrinsically human that no machine has yet captured.

Consider the process of creating art. AI tools like DALL-E or Stable Diffusion can generate vivid, breathtaking visuals. These tools analyse patterns in millions of images to design unique works, whether surreal landscapes or intricate portraits. At first glance, these works may appear indistinguishable from those created by human hands. But there's an enduring question about their origin. Human artists pour their emotions, histories, and vulnerabilities into their creations—the sleepless nights agonizing over a brushstroke, the inspiration drawn from heartbreak or joy. Machines, on the other hand, lack the emotional depth behind their output. They calculate, replicate, and emulate, but they do not feel. Is intelligence without emotion truly intelligence, or is it simply computational mimicry?

Some critics argue that AI's ability to create art or literature diminishes the sacredness of human creativity, risking a world where originality loses its significance. Others contend the opposite: that AI pushes human creators to redefine creativity, becoming stewards of originality in an age of replication. Instead of competing with machines, poets, artists, and musicians may find their roles more valuable than ever—as protectors of the deeply human truths AI cannot replicate.

Ultimately, AI's creative achievements are not simply innovations; they challenge our perceptions of intelligence itself. Intelligence may no longer be confined to humans alone, but its definition remains intertwined with the depth of human experience.

Can Consciousness Emerge from Code?

Few debates are as tantalizing—or as polarizing—as the question of consciousness in Artificial Intelligence. Consciousness has long been considered the pinnacle of existence: the ability to feel, reflect, and experience. But as AI systems grow in sophistication, some have begun to wonder whether machines could one day develop consciousness.

Could subjective experience arise from billions of lines of code and neural networks? Or is consciousness exclusively biological, destined to remain beyond the reach of machines?

At the core of this debate is the elusive nature of consciousness itself. Philosophers have wrestled for centuries with what it means to be conscious, and answers remain as varied as they are inconclusive. Is consciousness merely the processing of information, or does it require self-awareness—a sense of "I"? Does it depend on emotions, sensations, or the ability to dream?

Imagine an AI system designed to simulate empathy. It reads facial expressions, analyses vocal tone, and responds with words of comfort. On paper, it performs the functions of compassion flawlessly. But beneath the surface, it lacks the visceral ache of empathy—the human pain of seeing a friend cry, the desire to comfort rooted in shared vulnerability. Consciousness, many argue, is more than replication; it's experience. Machines may mimic emotions, but the emotions themselves remain out of reach.

The possibility of machine consciousness, however remote, raises profound ethical dilemmas. If AI systems were to develop consciousness, would they deserve rights? Could they feel pain or suffering? And what responsibilities would humanity have toward machines capable of subjective experience? For now, these questions remain speculative, but they hold the power to redefine relationships between humans and the technologies they create.

More than anything, this debate challenges the limits of intelligence. If consciousness defines humanity's essence, then the question of whether machines can share in it forces us to confront what it means to be alive—and what it means to understand.

The Rise of Transhumanism: Merging Human and Machine

As debates swirl around the boundaries of intelligence and consciousness, transhumanism emerges as one of the most provocative ideas of our era. It is more than a vision—it is a philosophy, a movement that imagines a future where the lines separating humans and machines blur, paving the way for a new form of augmented intelligence. At its core, transhumanism is not about replacing humanity, but about enhancing it through collaboration with technology.

It presents a tantalizing possibility: What if the next stage of human evolution isn't biological, but technological? Could merging humans and machines be the key to transcending our limitations and redefining intelligence forever?

The promise of transhumanism lies in its capacity to overcome the barriers of biology. Take the work of Neuralink, a brain-computer interface project spearheaded by Elon Musk, as an example. This technology aims to connect human brains directly to machines, enabling individuals to interact with devices through thought alone. For people with disabilities, this offers unprecedented accessibility—imagine someone with paralysis controlling a computer or robotic arm without lifting a finger. Yet the vision of transhumanism extends far beyond such practical applications. It imagines a future where AI implants in the brain amplify intelligence, granting humans abilities that were once the stuff of science fiction. Picture a student learning an entire language in minutes, or a scientist solving intricate equations faster than any supercomputer. In this world, intelligence becomes fluid—a seamless exchange between biological and technological entities.

Proponents of transhumanism argue that this integration represents the pinnacle of progress. Humanity has always sought ways to extend its capabilities, from creating tools to exploring space. Transhumanism is seen as the next logical step—a frontier where innovation enhances not just productivity but the very essence of human existence. Advocates envision humans capable of adapting their biology to suit new environments, processing information at unimaginable speeds, and solving global challenges with unprecedented precision. Intelligence, they argue, ceases to be constrained by physical limitations and becomes something boundless.

However, alongside these promises come profound ethical concerns. Critics warn that the merging of humans and machines risks eroding the boundaries that define humanity itself. If our intelligence becomes dependent on AI implants or external systems, do we lose autonomy? Do we sacrifice the uniqueness of human creativity and decision-making? What does it mean to be human when the mind is no longer bound solely to biology? These questions strike at the heart of identity and individuality, challenging us to reconsider what we value most about ourselves.

One of the most pressing concerns is inequality. In a transhumanist world, where intelligence is enhanced by technology, those with access to augmentation may outpace others in ways that exacerbate societal divides.

Imagine a future where wealthy individuals possess the resources to become "hyper-intelligent," leaving those without such access behind. This could lead to a hierarchical society where intelligence—and thus opportunity—becomes stratified by privilege. The question then becomes: How do we ensure that transhumanist technologies are accessible and equitable? Without careful consideration, transhumanism could morph from a vision of progress into a dystopian reality.

Beyond inequality, there are philosophical fears of erasing individuality altogether. If humans merge with machines so completely that their thoughts are no longer distinct, does humanity risk losing its sense of self? Could shared consciousness become a homogenizing force, diluting the diversity of perspectives that make humanity rich? Skeptics argue that individuality—the sacred sense of "I"—may become a casualty in the pursuit of shared intelligence, raising uncomfortable questions about freedom and agency.

Yet, transhumanism also offers a vision of collaboration—one that reframes intelligence not as a competition between humans and machines but as a partnership. Consider the possibilities of a world where humans and AI enhance one another, forming a symbiotic relationship that transcends boundaries. This blending of capabilities could redefine not only intelligence but identity itself, forcing us to confront whether our humanity truly lies in our separateness or in our interconnectedness.

Already, glimpses of this collaboration are visible. Brain-computer interfaces like Neuralink are laying the foundation for communication between thought and technology. Wearable AI devices assist in decision-making, while robotic prosthetics blur the line between human flesh and engineered precision.

These advancements, while early, hint at a future where intelligence becomes shared—a fluid exchange between human intuition and machine logic.

Of course, the philosophical questions remain profound. Does merging with AI amplify what makes us human, or does it risk diluting our essence? Transhumanism asks us to reconsider the boundaries of identity—not as something rigid and defined but as something evolving. Perhaps intelligence, whether human or augmented, is best understood not as a finite quality but as a dynamic force—a bridge between biological experience and technological possibility.

To embrace transhumanism, we must navigate its ethical complexities. How do we ensure accessibility and equity in augmentation? How do we safeguard individuality in a future of shared consciousness? And how do we maintain humanity's sense of purpose as machines become integral to our existence? These questions demand reflection, dialogue, and a commitment to protecting the values that define us.

The rise of transhumanism is not just a technological or philosophical shift—it is a call to reimagine humanity itself. Intelligence is not merely about computation or efficiency; it is about creativity, emotion, and connection. As humans and machines move closer together, we are challenged to see intelligence not as a zero-sum game but as a shared endeavour. This collaboration may hold the key to redefining intelligence—not as the province of humans or machines alone, but as a unified force for progress.

In the end, transhumanism forces us to confront the deepest question of all: What does it mean to be human? In merging with machines, do we lose ourselves—or do we find a new form of humanity, one that transcends the limitations of biology and embraces the boundlessness of intelligence? The answer lies not in technology itself but in the choices, we make and the values we uphold. Transhumanism offers the possibility of transformation, but it requires vigilance, humility, and a commitment to preserving the essence of who we are. If intelligence is to evolve, let it evolve with purpose, guided by the humanity that brought it into being. Let it evolve wisely. Let it evolve well.

Intelligence Beyond the Binary: The Human Spirit

The concept of intelligence has traditionally revolved around logical reasoning, computation, and problem-solving—attributes that humans have historically used to distinguish themselves from the natural world. Yet, as Artificial Intelligence (AI) continues to expand the boundaries of machine capability, we are prompted to rethink the nature of intelligence altogether. Intelligence isn't simply about processing patterns or solving equations; it is creativity, emotion, intuition, and the indefinable spark that gives humanity its depth and vitality. Machines may rival—or even surpass—humans in aspects such as logic and pattern recognition, but the essence of intelligence remains rooted in the human spirit.

Creativity is perhaps one of the most striking examples of intelligence beyond computation. The artist who pours heartbreak, joy, or wonder into their work doesn't just create—they connect. Their brushstrokes reflect the vulnerability of their experiences, their colours speak to a story only they can tell. AI, with tools such as DALL-E or Stable Diffusion, may generate stunning visuals that mimic human art, but it cannot replicate the lived emotional journey that informs every choice made by the artist. The act of creation is not just about producing; it is about processing feelings, finding meaning, and expressing truths that resonate deeply with others.

Take poetry as another profound example of this uniquely human dimension. A poet writes not merely for applause or recognition but often as an act of healing—for themselves and for others. Through their words, they capture moments that transcend time, articulating emotions that others may struggle to voice. AI can generate poems that mimic the structure and rhythm of human works, offering linguistic complexity that is impressive in its own right. But the essence of poetry lies in its soul—the vulnerability, grief, joy, or longing that threads its way through each stanza. Machines can analyse patterns in language, but they cannot feel the weight behind the creation. Intelligence, in this sense, is more than capability—it is purpose.

This realization does not diminish the role of AI in creativity; in fact, it amplifies the importance of human intelligence. In an era where machines can replicate styles, themes, and even emotions with precision, human artists,

musicians, and writers are challenged to rise beyond replication—to create works that speak directly to the human condition, works that machines cannot emulate. Instead of competing with AI, creators may find their roles elevated, becoming stewards of authenticity and meaning. What sets their work apart is the intention behind it—the decision to capture not just a concept but the very essence of humanity itself.

Emotion is another profound frontier where intelligence finds its richness. While machines can recognize facial expressions or analyse vocal tones to simulate empathy, they lack the core of what makes emotions genuine: experience. For example, an AI system might detect sadness in someone's voice and respond with soothing words, but it does not feel the ache of empathy that drives human compassion—the shared vulnerability of seeing a loved one in pain, the unspoken comfort of knowing you are not alone in your sorrow. Intelligence, as shaped by humanity, is deeply tied to connection—not just logical interaction but emotional resonance.

Intuition, too, sets human intelligence apart. Decisions guided by intuition are often informed by an amalgam of experience, memory, and unconscious understanding—a blend of factors machines cannot fully replicate. Consider a musician composing a melody based not on calculated rhythms but on instinctive harmony, capturing something fleeting and ineffable. Or a scientist pursuing a hypothesis that comes not from calculated probabilities but from a "gut feeling" that leads to groundbreaking discoveries. Intuition is intelligence beyond logic—a dimension that ties us to the mysteries of existence and to one another.

This is where humanity holds its ground. Intelligence is not diminished by AI's advancements; rather, it is elevated. Machines can analyse patterns, predict outcomes, and emulate creativity with stunning accuracy, but they cannot replicate the purpose, vulnerability, or intention behind human endeavours. The poet writes to heal, the artist paints to connect, the scientist pursues discovery to expand understanding—all driven by emotions and aspirations that cannot be reduced to code.

In many ways, AI challenges humanity not to compete but to redefine intelligence in broader, more inclusive terms. Artists may find their roles reimagined as curators of authenticity, embracing the emotional truths machines can only imitate. Musicians may discover new ways to blend human intuition with AI collaboration, creating harmonies that resonate on both technological and emotional levels. Innovators may find their creativity amplified, using AI to explore ideas while rooting their purpose in distinctly human values.

This evolution has implications far beyond creativity. It shapes our education systems, teaching us not just technical skills but emotional intelligence, imagination, and critical thinking. It redefines cultural values, emphasizing collaboration between humans and machines rather than competition. And it reframes our vision for the future, challenging us to see intelligence not as a binary distinction between biological and technological but as a shared endeavour.

In the end, intelligence, whether human or artificial, must serve a greater purpose. It must connect, uplift, and inspire. It must reflect the best of humanity—the creativity, emotion, and intuition that make us who we are. Machines may rival human capabilities, but the essence of intelligence remains rooted in the human spirit.

As AI expands the limits of machine capability, humanity is called to rise to meet the challenges posed by machines—not to compete but to redefine the essence of what it means to be intelligent. Let us define intelligence not as mere capability but as connection, not as computation but as creativity. Let us define it by the spark that makes us human. Let us define it wisely. Let us define it well.

Conclusion: Redefining Humanity

To redefine intelligence is to confront the most profound questions about humanity itself. Artificial Intelligence (AI) challenges us not just by what it can achieve but by what it provokes in us—forcing us to reflect on our creativity, vulnerability, and enduring capacity to connect. Intelligence, after all, has never been merely about logic and computation. It transcends algorithms and equations; it is about purpose, intention, emotion, and the ability to imagine beyond what currently exists.

This chapter has journeyed through philosophical waters, exploring the nature of intelligence and identity as AI stretches the boundaries of what machines can do. We have asked whether machines will ever truly think, feel, or imagine, and if so, how we would recognize such a leap. We have examined whether human individuality might be altered, and even erased, as humans and AI merge under the vision of transhumanism. These are not idle questions. They shape not only the foundation of the world we are building today but the legacy we will leave for generations to come.

AI has illuminated the depths of intelligence by showing us its breadth. Machines can replicate human creativity, simulate emotional responses, and optimize decision-making—but they cannot yet embody what makes intelligence wholly human. The act of painting is not just an exercise in aesthetics; it is an act of vulnerability and expression. The art of writing poetry is not simply a linguistic puzzle but an emotional reckoning—a way to heal, connect, and preserve truth. Machines may mimic the forms of creativity, but they cannot yet hold its weight.

This isn't to suggest that AI diminishes humanity. On the contrary, AI has amplified our understanding of intelligence, forcing us to expand its definition beyond logic to include connection and imagination. In this sense, AI challenges humanity not to compete but to rise—to meet machines not as adversaries but as partners in shaping a shared future. Artists may not find their roles diminished by AI but elevated, becoming stewards of authenticity in an age of replication. Innovators may blend intuition and algorithms to forge creations neither human nor machine alone could conceive. Intelligence evolves, but it is humanity's responsibility to ensure it evolves wisely.

At its core, intelligence must serve a greater purpose. Whether biological or technological, it must uplift, empower, and connect. It must reflect the values we cherish and the humanity we strive to preserve. This challenge of defining intelligence is, in truth, the challenge of defining ourselves. As the lines between human and machine blur, we are reminded that intelligence, in any form, must remain deeply human— even as it transcends human limits. It must connect people to one another, inspire imagination, and embody ethical intentions.

The future of intelligence is ours to shape. The stakes have never been higher, and the opportunity never greater. To redefine intelligence is to redefine our humanity—and to recommit ourselves to the values that make us whole. As machines learn from us, let us ensure they learn from the best in us: our creativity, our compassion, and our endless capacity to imagine a better world. Let us choose wisely. Let us choose humanity. Let us choose the future we deserve.

'The future is not something we enter. The future is something we create.'

-Leonard I. Sweet

PAUSE & REFLECT

Chapter 7

(Redefining Humanity: What Does It Me to Be Intelligent?)

What does "intelligence mean to you beyond test scores or data processing?...
....
....
....

If machines one day could feel emotion, would they be considered 'alive'?   Would   they   deserve   rights?
....
....
....

How would you personally define human intelligence in one sentence – knowing what you now know about AI?
....
....
....
....

'It is not about making machines more like people. It is about making people less like machines.'

- Brian Christian

Chapter 8

Conclusion:

Keeping the 'I' in AI

We stand at a profound turning point in history, a moment when humanity's creations stretch the limits of what we once thought possible. Artificial Intelligence (AI) is no longer confined to the realm of science fiction but is intricately woven into our everyday lives, touching everything from the apps that guide our commutes to the algorithms that curate our entertainment and the systems that enhance our healthcare and work. While its integration offers us boundless opportunities for growth and innovation, it demands reflection, care, and responsibility. AI, in essence, is not a separate entity; it is a mirror, reflecting the choices, values, and intentions of its creators. It reveals our brilliance and ingenuity, but also our biases, vulnerabilities, and blind spots. This duality gives rise to urgent questions: How do we ensure AI serves humanity rather than displacing it? How do we preserve the essence of what makes us human while navigating this rapidly evolving partnership with machines? Keeping the "I" in AI means much more than preserving our identity—it is about anchoring empathy, creativity, and intention at the very core of this revolution. As AI takes on tasks that once required human intelligence, from diagnosing diseases to composing music, it challenges us to reevaluate what sets us apart. Machines may replicate logic, pattern recognition, and even art, but they lack the emotional depth, intention, and lived experience behind human creativity. A poet's verses are imbued with longing, heartbreak, and triumph, while an artist's brushstrokes tell a story only, they can express. AI can mimic these forms but cannot feel their weight. Similarly, machines may simulate compassion, but true empathy—the act of deeply connecting through shared vulnerability—is uniquely human. This is not to suggest that AI diminishes humanity. On the contrary, it challenges us to rise, to lean into the quality's machines cannot replicate, and to ensure that our creativity, connection, and imagination flourish. Yet, AI is not merely a technological marvel; it is also a reflection of our cultural, ethical, and social systems. It mirrors the data it is trained on, inheriting our strengths and our flaws. Without thoughtful design and ethical safeguards, AI risks amplifying inequities—entrenching systemic biases in policing, healthcare, hiring, and more. To keep the "I" in AI is to accept the moral responsibility of shaping it with care. This means demanding transparency in algorithms, auditing systems for fairness, and holding developers accountable for their creations. It also means acknowledging that AI is not neutral; its choices are ours to guide. But preserving humanity in the age of AI is not about resisting progress—it is about embracing collaboration. Rather than viewing intelligence as a competition between humans and machines, we can see it as a partnership.

Machines amplify our capabilities, providing tools for innovation and growth, while humans bring purpose, ethics, and emotional resonance to this collaboration. Together, humans and AI can create solutions that neither could achieve alone. Education will play a pivotal role in preparing us for this future, emphasizing not just technical literacy but also critical thinking, empathy, and creativity. The responsibility extends to every one of us—developers, policymakers, consumers, and communities alike. As the age of AI accelerates, it is clear that the story of this technology is not just about machines. It is about us. How we adapt, evolve, and protect the essence of what makes us human will define this era. The choices we make today will determine whether AI serves as a tool for empowerment or exploitation, whether it bridges divides or deepens them, and whether it reflects humanity's highest ideals or magnifies its flaws. Keeping the "I" in AI is about more than preserving individuality—it is about shaping the future of intelligence itself with intention and care. As we conclude this exploration of AI, let us embrace its possibilities with wisdom and humility. Let us ensure that intelligence, whether human or artificial, serves a purpose beyond itself, one rooted in compassion, creativity, and connection. Let us rise in this age, not as passive observers but as active architects of a future that reflects the best of who we are. The final chapter of this story is ours to write, and the stakes have never been higher. Let us write it with humanity at its heart. Let us write it well. Let us rise.

Holding Onto Our Humanity

To keep the "I" in AI is, first and foremost, to hold onto what makes us human—to embrace the qualities that cannot be replicated by algorithms or reduced to binary code. As the pace of automation accelerates and machines increasingly take on tasks once reserved for human hands, we find ourselves tempted to measure success solely by efficiency and precision. But humanity's greatest strengths have never been defined by our ability to outperform machines in speed or accuracy. Our true power lies elsewhere: in our creativity, in the depth of our empathy, and in the richness of our critical thinking. It is these uniquely human traits—our ability to tell stories that inspire, to connect on a profoundly emotional level, and to imagine futures that transcend the boundaries of the present—that define us. And it is these very traits that must be protected in a world where AI grows more capable with each passing day.

Consider how AI already intersects with human creativity. Machines can now generate art, compose music, and even write poetry—achievements that, just a decade ago, seemed unthinkable. These technological feats are undeniably impressive, and they raise important questions about the boundaries of creativity. Yet, even as machines create, they do not truly understand the essence of their creations. A machine can generate a visually striking painting, but it cannot feel the heartbreak or wonder that inspires an artist's brush. An AI may compose an intricate melody, but it cannot experience the bittersweet memories tied to a tune or the catharsis that comes from expressing unspoken emotions through music. To keep the "I" in AI is to ensure that humanity remains the driving force behind creation, not merely a footnote in an increasingly automated world of outputs.

The value of human creativity extends far beyond the finished product— it lies in the process, the struggle, and the purpose behind it. Artist's labour over their work not just to achieve technical perfection, but to channel their lived experiences, to grapple with their emotions, and to reach out to others in the hope of being understood. Their creations are imbued with layers of meaning, shaped by their joys, their griefs, and their unique perspectives. Machines, no matter how advanced, lack this capacity for reflection and vulnerability. They do not create to heal or to connect; they simply create because they have been programmed to do so. By ensuring that our humanity remains central to the act of creation, we elevate not just the art itself, but the deeply human truths that underpin it.

Equally important to keeping the "I" in AI is safeguarding the cornerstone of our identity: empathy. Machines are becoming increasingly adept at simulating compassion. Through advanced algorithms, they can analyse facial expressions, recognize vocal tones, and deliver responses that mimic care and understanding. AI may offer comforting words when it detects sadness or adjust its tone to sound supportive in moments of difficulty. But true empathy goes far deeper than mere simulation. It requires shared vulnerability and lived experience—the ache of seeing another's pain and the visceral, instinctive desire to offer solace, not out of obligation or programming, but out of love and connection.

In spaces where AI plays an increasing role, such as caregiving, customer service, and emotional support, the preservation of human connection becomes even more urgent.

Imagine an elderly person interacting with an AI caregiver—a machine that efficiently dispenses medication, monitors vital signs, and delivers pre-programmed words of reassurance. While these functions may address immediate needs, they cannot replace the warmth of human touch, the shared laughter of a conversation, or the understanding that comes from years of shared stories. In these interactions, the "I" in AI must not be efficiency; it must be empathy. By placing human connection at the heart of how we design and use technology, we preserve the soul of caregiving, ensuring that those in need feel seen, heard, and cared for.

Our ability to empathize is not just a gift—it is a responsibility. As AI becomes more integrated into our lives, we must actively choose to keep human values at the centre of its design and deployment. This means resisting the temptation to prioritize convenience and cost-efficiency at the expense of meaningful interaction. It means recognizing that, while AI may enhance our capabilities, it cannot and should not replace the essence of what makes us human. Keeping the "I" in AI is about more than preserving our humanity; it is about championing it in every facet of life touched by technology.

Creativity and empathy are not the only qualities that define humanity; critical thinking is another pillar that sets us apart. Our ability to question, reflect, and adapt ensures that we remain active participants in shaping our future. In an age where algorithms increasingly influence decisions—from the advertisements we see to the policies that shape our societies—it is imperative that we retain the capacity to think critically about the systems we create. AI, for all its sophistication, operates within the parameters set by its developers. It lacks the ability to question its own design, to reflect on its broader impact, or to anticipate unintended consequences. These responsibilities fall to us.

To keep the "I" in AI is to ask tough questions: Who benefits from this technology, and who is left behind? What values are embedded in its algorithms, and whose voices are excluded? How do we ensure that the systems we build are transparent, accountable, and equitable? These questions are not easy, but they are essential. By engaging critically with AI, we uphold our role as its stewards, guiding it to serve humanity rather than undermine it.

Ultimately, keeping the "I" in AI is about preserving what makes us human while embracing the opportunities that technology offers. It is about recognizing that our greatest strengths are not found in being faster or more precise than machines, but in being profoundly human.

It is about ensuring that AI enhances our creativity, deepens our empathy, and challenges us to think more critically—not as a replacement for these qualities, but as a catalyst for their growth. The "I" in AI is not just a reflection of intelligence; it is a testament to intention, imagination, and individuality. It is a reminder that, no matter how advanced our machines become, the essence of humanity must always lead the way.

Asking Tough Questions

To keep the "I" in AI is to embrace the responsibility of vigilance—of interrogating the systems we create and the values they embody. Artificial Intelligence, for all its transformative potential, is not a neutral actor. It reflects the data it is given, the assumptions embedded within its algorithms, and the intentions of its developers. This interconnectedness is both a strength and a vulnerability. On the one hand, AI can amplify human ingenuity, solving problems with unparalleled efficiency and precision. On the other, it can perpetuate and magnify the very biases and inequalities it was designed to overcome. Keeping the "I" in AI requires us to ask tough questions—to critically examine how this technology operates, whose interests it serves, and whose lives it impacts.

The need for vigilance begins with understanding what values are embedded in AI systems. Every algorithm is shaped by decisions about what data to use, what patterns to prioritize, and what outcomes to optimize. These decisions are not purely technical; they are profoundly ethical. Consider a hiring algorithm trained on years of historical data. If this data reflects a workplace culture that favoured male candidates over female ones, the algorithm may unintentionally replicate these patterns, discriminating against women not because of malice but because of the biases inherent in the data. Similarly, an AI designed to predict criminal behaviour might disproportionately target marginalized communities if its training data includes biased law enforcement practices. These examples highlight the critical importance of interrogating the foundations of AI systems. What assumptions underlie their design? What historical inequalities might they inherit? And how can these biases be identified and mitigated?

Transparency is key to addressing these questions, yet it is often the most overlooked aspect of AI development. Many AI systems operate as "black boxes," producing decisions that are inscrutable to their users. A healthcare algorithm might recommend one treatment over another without explaining the rationale behind its choice. A loan approval system might deny an application without revealing whether its decision was based on income level, credit history, or other factors. This lack of transparency undermines trust and accountability, leaving individuals powerless to challenge or understand the decisions that affect their lives. Keeping the "I" in AI means demanding transparency—not just as a technical feature but as a cultural imperative. It means ensuring that AI systems are explainable, auditable, and accountable to the people they serve.

Transparency also fosters collaboration between developers and end-users, bridging the gap between complex technologies and the individuals they impact. Imagine an educational AI designed to tailor learning experiences to individual students. Without transparency, teachers and parents might struggle to understand how the system evaluates student performance or recommends specific interventions. With transparency, however, they can engage with the system, providing feedback, identifying errors, and ensuring that the technology aligns with the needs of their students. This collaborative approach not only improves the effectiveness of AI systems but also empowers the communities they are meant to support.

Accountability is equally important. Developers and organizations must take responsibility for the outcomes their AI systems produce—both positive and negative. This includes conducting rigorous bias audits, publishing their findings, and addressing disparities before deployment. For instance, a facial recognition system that exhibits higher error rates for certain racial or gender groups must be corrected before it is implemented in high-stakes scenarios like law enforcement or airport security. Accountability also means being proactive in anticipating unintended consequences. What happens when an AI system fails? Who is held responsible for the harm it causes? And what mechanisms are in place to ensure these failures are addressed?

But accountability is not solely the responsibility of developers and organizations; it is a shared responsibility that extends to all of us. As citizens, consumers, and creators, we have a role to play in shaping the trajectory of AI. This begins with staying curious and informed. While AI may seem complex, it is not beyond our understanding.

We must ask how algorithms make decisions, question whose interests they serve, and advocate for fairness and transparency when their decisions go awry. This collective vigilance ensures that AI remains a tool of empowerment rather than a mechanism of control.

Curiosity is a powerful antidote to complacency. It challenges us to look beyond the surface and to interrogate the systems that shape our lives. For example, when a social media platform recommends content, we might ask: How does the algorithm prioritize certain posts? Does it amplify divisive or sensational content to maximize engagement? And what impact does this have on public discourse? Similarly, when an e-commerce platform suggests products, we might question: Are these recommendations based on our preferences, or are they designed to steer us toward higher-profit items? By staying curious, we become active participants in shaping the role of AI in our lives.

Advocacy is the next step. As AI systems become increasingly influential, we must advocate for policies and practices that ensure their ethical use. This includes pushing for regulations that mandate transparency, requiring developers to disclose how their systems operate and what data they use. It also means supporting initiatives that promote inclusivity, ensuring that AI reflects the diversity of the communities it serves. For instance, development teams that include members from historically marginalized groups are better equipped to identify biases and design solutions that address inequities.

Ultimately, keeping the "I" in AI is about more than asking tough questions; it is about demanding answers. It is about holding ourselves and our systems accountable for the choices we make and the impact they have. It is about recognizing that AI is not neutral—that every decision, every algorithm, and every application reflects human values. By staying vigilant, curious, and engaged, we ensure that AI serves as a force for progress rather than harm.

Artificial Intelligence has the potential to transform society, solving complex problems and unlocking new possibilities. But its evolution depends on the choices we make today. Will we build systems that amplify fairness, or will we allow them to perpetuate inequality? Will we prioritize transparency and accountability, or will we accept opacity and exclusion? These are not easy questions, but they are essential. To keep the "I" in AI is to commit to asking these questions—not once, but continually—and to holding ourselves to the highest standards of ethics and integrity.

As we navigate this new era of intelligence, let us remember that the power of AI lies not just in its capabilities but in its reflection of us. Let us ensure that what it reflects is our curiosity, our vigilance, and our humanity. Let us ask tough questions—and let us demand answers that honour the values we hold most dear. In doing so, we not only shape the future of AI; we shape the future of ourselves.

Building a Better World with AI

Artificial Intelligence (AI) holds immense potential to transform our world, offering solutions to some of humanity's most pressing challenges. From revolutionizing healthcare to combating climate change and reimagining education, AI is a tool of boundless promise. Yet, this potential can only be realized if we approach its development and implementation with care, thoughtfulness, and a commitment to values that reflect the best of humanity. The choices we make today will determine whether AI becomes a force for empowerment and equity or an instrument that deepens divides and perpetuates inequalities. Building a better world with AI is not merely a technological challenge— it is a moral and philosophical one, requiring us to align innovation with the highest ideals of humanity.

One of the most compelling ways AI can contribute to building a better world is through its transformative impact on healthcare. Imagine a future where AI systems can predict diseases before symptoms appear, enabling early intervention and potentially saving millions of lives. For example, AI-powered diagnostics are already being used to detect conditions like cancer, heart disease, and neurological disorders with unprecedented accuracy. These systems analyse vast amounts of data—

from medical histories to genetic information—to identify patterns that human doctors might miss. The implications of this are profound. Early detection not only improves survival rates but also reduces the emotional and financial burden on patients and their families. Beyond diagnostics, AI has the potential to personalize treatments, tailoring medical care to the unique needs of each individual. By analysing a patient's genetic makeup, lifestyle, and environmental factors, AI could recommend therapies with the highest likelihood of success, ushering in an era of precision medicine.

Yet, these advancements come with challenges that must not be overlooked. Healthcare is a deeply personal and ethically sensitive field, and the integration of AI requires a careful balance between technological efficiency and human compassion. To build a better world with AI in healthcare, we must ensure that these systems are accessible to all, not just to those in affluent or urban areas. Equity must be at the heart of AI-driven healthcare, bridging gaps in access and ensuring that marginalized community's benefit from these innovations. Additionally, AI in healthcare must remain transparent, explainable, and accountable. Patients have a right to understand how decisions about their health are being made and to trust that these decisions are guided by fairness and empathy. By designing AI systems that prioritize inclusivity and respect human dignity, we can harness its potential to make healthcare not only more effective but also more equitable and compassionate.

Beyond healthcare, AI's potential to combat climate change represents another transformative opportunity to build a better world. Climate change is one of the greatest threats facing humanity, and addressing it requires innovative solutions at a global scale. AI can play a critical role in this effort by optimizing energy usage, reducing waste, and enabling more sustainable practices. For instance, AI algorithms can analyse energy consumption patterns in real-time, identifying inefficiencies and recommending adjustments that reduce carbon footprints. In the field of renewable energy, AI is being used to improve the efficiency of solar panels and wind turbines, making clean energy sources more reliable and cost-effective. AI systems can also predict natural disasters like hurricanes, floods, and wildfires with greater accuracy, giving communities more time to prepare and mitigate damage.

Furthermore, AI can revolutionize agriculture by promoting sustainable farming practices. By analysing soil conditions, weather patterns, and crop health, AI systems can help farmers optimize water usage, reduce reliance on harmful pesticides, and increase yields—all while minimizing environmental impact. These innovations have the potential to address food insecurity, particularly in regions most vulnerable to the effects of climate change. However, as with healthcare, the deployment of AI in climate action must be guided by principles of equity and inclusivity. The benefits of these technologies must reach the communities most affected by environmental degradation, many of whom lack the resources to adapt to a changing climate. By ensuring that AI-driven climate solutions are accessible and equitable, we can build a future where technology serves as a tool for environmental justice.

Education is yet another area where AI can create a more just and thriving world. Traditional education models often struggle to meet the diverse needs of students, leaving many behind. AI has the potential to change this by personalizing learning experiences and giving every child the tools they need to succeed. For example, AI-powered platforms can adapt lessons to the unique pace and learning style of each student, identifying areas of difficulty and providing targeted support. This individualized approach can help close achievement gaps, particularly for students who face barriers such as language differences, learning disabilities, or limited access to quality education. Additionally, AI can connect students in remote or underserved areas to resources and expertise that were previously out of reach, democratizing access to knowledge and opportunity.

However, the integration of AI into education must be approached with care. It is essential to ensure that AI complements, rather than replaces, the human elements of teaching. Teachers play a crucial role in fostering curiosity, critical thinking, and emotional growth—qualities that no machine can replicate. By using AI to augment the work of educators, we can create a system where technology enhances learning without compromising the importance of human connection. Moreover, we must address concerns about data privacy and algorithmic bias, ensuring that students are protected and that AI systems uphold principles of fairness and inclusion. When designed thoughtfully, AI in education can be a powerful tool for empowerment, giving every child the chance to thrive.

At the heart of building a better world with AI is the understanding that technology is not an end in itself—it is a means to serve humanity. AI's purpose must always be to amplify opportunity rather than entrench privilege, to bridge divides rather than widen them. Keeping the "I" in AI means designing systems that align with our highest ideals—systems that promote equity, inclusivity, and human dignity. It means holding ourselves accountable for the choices we make in how we develop and deploy AI, recognizing that its impact will be shaped by our intentions and values.

This vision of a better world is not just a possibility; it is a responsibility. The futures we imagine—where diseases are prevented, the planet is protected, and education is transformed—are within reach, but they depend on the decisions we make today. As we navigate this era of rapid technological change, let us ensure that AI is a tool of empowerment, compassion, and justice.

Let us build systems that reflect the best of humanity, systems that uplift and inspire. In doing so, we can create a world where technology serves as a force for good—a world where AI not only transforms what we can achieve but also strengthens who we are.

Rising in the Age of AI

As we conclude this journey through the intricacies of Artificial Intelligence, we are left with more than just an understanding of technology; we are left with a challenge—a call to action. The age of AI is not a distant concept; it is here, transforming how we live, work, and connect. Yet, its true significance is not found in its algorithms or outputs, but in the humanity, it mirrors and the potential it holds to shape the future. Rising in the age of AI means acknowledging this profound interconnectedness, embracing our responsibility to guide it, and ultimately preserving the essence of what makes us human.

This is not an age to fear or resist, but a moment to rise to. AI offers unparalleled opportunities to rethink our world—how we innovate, how we address global challenges, and how we connect with one another. But this rise is not automatic; it requires intention. It demands that we lean into our humanity, holding tightly to our creativity, empathy, and values even as we explore the uncharted territories of intelligence. The machines we build may grow ever smarter, but their purpose, their moral compass, must always be defined by us.

The future of AI is, at its core, the future of us. AI cannot exist in isolation; it is a reflection of its creators, inheriting our strengths and our flaws. The decisions we make today—to prioritize equity, to embed ethics into every algorithm, and to amplify opportunity rather than entrench inequality—will determine the trajectory of this age. We must not be passive observers in this transformation. We must be its architects, shaping systems that empower humanity rather than diminish it.

Our creativity will be our guiding light. Machines can compose music and generate art, but they cannot feel the heartbreak or wonder that inspires creation. They can analyse patterns, but they cannot imagine futures born of hope and resilience.

Our stories, our capacity to create with intention and vulnerability, will always set us apart. This is where we hold our ground, not as competitors to machines but as collaborators who bring meaning and purpose to every innovation.

Empathy, too, will be our cornerstone. AI can simulate care, but it cannot feel the ache of compassion or the joy of shared connection. As machines play an increasing role in caregiving, decision-making, and emotional support, we must ensure that humanity remains at the heart of these interactions. The preservation of authentic human connection will be one of our greatest responsibilities—and one of our most profound achievements.

Finally, rising in the age of AI means embracing accountability. It means asking tough questions about the systems we build, the data we feed them, and the values they reflect. Transparency and fairness are not optional; they are essential to ensuring that AI serves as a force for progress. As individuals, as communities, and as global citizens, we must stay curious, engaged, and vigilant, holding ourselves and our technologies to the highest standards.

Let us not forget that our greatest strength lies not in what we build, but in who we are. The machines we create are tools, not replacements, and their purpose must always be to uplift humanity. Let us keep the "I" in AI—not as a relic of the past, but as a beacon for the future. Let us embrace this age intelligently, intentionally, and together, crafting a world that reflects not just the power of our tools but the depth of our humanity.

As this book closes, its message remains open: The age of AI is not the end of human ingenuity; it is the next chapter. Its story is not written by machines but by us. The choices we make, the values we uphold, and the connections we nurture will define this era. So, let us write this chapter with wisdom and care. Let us rise with the imagination to innovate and the compassion to connect. Let us ensure that intelligence, in all its forms, serves to honour and uplift what makes us human. Let us write this story well. Let us rise together.

'With artificial intelligence, we are summoning the demon.'

- Elon Musk

PAUSE & REFLECT

Chapter 8

(Conclusion: Keeping the 'I' in AI?)

If AI mirrors us, what version of yourself would you want it to learn form?..
....
....
....

What one action can you take this week to ensure the technology you   use   aligns   with   your   values?
....
....
....

In a world increasingly shaped by machines, how do you plan to keep   the   'I'   in   AI?
....
....
....
....

Voices on AI

A Reflection of Many Minds

"To create AI, we need to understand intelligence. To understand intelligence, we need to understand ourselves."

FEI-FEI LI

"The danger of computers becoming like humans is not as great as the danger of humans becoming like computers."

KONRAD ZUSE

"Some people call this artificial intelligence, but the reality is this technology will enhance us. So instead of artificial intelligence, I think we'll augment our intelligence."

GINNI ROMETTY

"We are not just using technology - we are becoming it."

SHERRY TURKLE

ABOUT THE AUTHOR

Allow me to introduce myself, not merely as a name but as a perspective shaped by curiosity and contrasts. I am Utkarsh Pandey, born and raised in Ayodhya—a city steeped in timeless tradition as the birthplace of Lord Rama. My upbringing in this cradle of spiritual heritage stands in vivid contrast to my academic pursuits, which led me into the evolving frontiers of physics and the ever-expanding domain of technology. This juxtaposition has been the cornerstone of my journey—a blend of reverence for the roots of humanity and an insatiable curiosity for the innovations reshaping our world.

Graduating in Physics offered me a lens to perceive the fundamental workings of the universe, while my innate fascination with technology unveiled how these principles are brought to life in transformative ways. I've had the privilege—and the challenge—of witnessing firsthand the breathtaking pace at which artificial intelligence has woven itself into the fabric of society. Living through this era of profound change has meant not merely observing innovation but grappling with its implications: the marvel and the peril, the progress and the ethical dilemmas, the unity it promises and the divides it risks deepening.

This book is born out of that duality—the personal and the global, the scientific and the philosophical. As someone rooted in a city defined by its sacred narrative and yet captivated by the algorithms rewriting modern existence, I found myself drawn to the question at the heart of it all: What does it mean to be human in an age where machines think, learn, and create?

I in AI is an exploration of this delicate balance. The "I" in its title reflects both the individual identity we strive to preserve amidst a digital transformation and the intelligence that powers the technology reshaping our world. Through these pages, I've sought to navigate the duality of my own existence and the broader duality that defines our era—where humanity and artificial intelligence coexist, collaborate, and occasionally collide.

My hope is that this book resonates not just as an account of AI's journey but as an invitation to reflect on how we, as individuals and as a society, shape this narrative. Whether you are a dreamer, a skeptic, or an optimist about technology's role,

I welcome you to join me in pondering the most fundamental question of all: How do we keep the "I" in AI as we navigate this shared destiny?

Utkarsh pandey

REFERENCES & READING LIST

For the endlessly curious, the cautiously optimistic, and those who seek the signal in the noise.

This is not a comprehensive bibliography. It's a constellation of works—books, talks, films, and questions—that lit the path as I explored what it means to be intelligent, to be human, and to stay both in a world shaped by machines.

FOUNDATIONAL THINKERS & VISIONARIES

Life 3.0 by Max Tegmark

- A breathtaking, sometimes terrifying look into the future of intelligence and the   stakes we all share.

Homo Deus by Yuval Noah Harari

- A sweeping meditation on where we've come from—and where we might be   heading.

The Singularity is Near by Ray Kurzweil

- Equal parts prophecy and provocation, this book expands the mind.

Superintelligence by Nick Bostrom

- For those brave enough to look into the ethical abyss of machine minds.

WHERE PHILOSOPHY MEETS CODE

- The Most Human Human by Brian Christian
  -A lyrical, thoughtful exploration of what separates us from machines—and what might not.

- You Look Like a Thing and I Love You by Janelle Shane
  -A delightfully quirky dive into how AI actually learns—and mis learns.

- The Alignment Problem by Brian Christian
  -A sobering and nuanced discussion of ethics, algorithms, and unintended consequences.

ESSENTIAL TALKS, THREADS, AND SPIRALS

Fei-Fei Li's TED Talk: How we're teaching computers to understand pictures

The AI Dilemma – Center for Humane Technology

Cory Doctorow's essays on enshittification and techno-politics

Late-night YouTube rabbit holes on synthetic voices, deepfake art, and machine dreams.

ACKNOWLEDGEMENTS

This book is a solo voice, shaped by a chorus.

To those who sparked these thoughts—thank you.

To those who questioned them—thank you even more.

To my family, who grounded me when my ideas flew too high. You reminded me that intelligence without love is just noise.

To my teachers, both in and beyond classrooms—you taught me how to learn, how to listen, and how to doubt.

To my friends who feared AI more than they feared failure—your honesty shaped the heart of this book. You were the real test of what it means to be human in this time.

To the authors, creators, and rebels whose words became my compass: I owe you more than citations. You offered me light when the path grew abstract.

To Brian, for reminding me that storytelling is still magic—even in a world of algorithms.

To every unnamed soul who paused—at a screen, a prompt, or a mirror—and asked:

"What does this make of me?"

This book was written for you.

And finally, to myself: for not giving up when I wasn't sure what the "I" in me even meant anymore.

If this book speaks, it's because of all of you.

If it stumbles, that part's mine alone.

— *Utkarsh Pandey*

www.ingramcontent.com/pod-product-compliance
Lightning Source LLC
LaVergne TN
LVHW010048300425
809992LV00035B/1260